CALVERT
MATH

CALVERT®
SCHOOL

Calvert Math is based upon a previously published textbook series. Calvert School has customized the textbooks using the mathematical principles developed by the original authors. Calvert School wishes to thank the authors for their cooperation. They are:

Audrey V. Buffington
Mathematics Teacher
Wayland Public Schools
Wayland, Massachusetts

Alice R. Garr
Mathematics Department Chairperson
Herricks Middle School
Albertson, New York

Jay Graening
Professor of Mathematics
and Secondary Education
University of Arkansas
Fayetteville, Arkansas

Philip P. Halloran
Professor, Mathematical Sciences
Central Connecticut State University
New Britain, Connecticut

Michael Mahaffey
Associate Professor,
Mathematics Education
University of Georgia
Athens, Georgia

Mary A. O'Neal
Mathematics Laboratory Teacher
Brentwood Unified Science
Magnet School
Los Angeles, California

John H. Stoeckinger
Mathematics Department Chairperson
Carmel High School
Carmel, Indiana

Glen Vannatta
Former Mathematics Supervisor
Special Mathematics Consultant
Indianapolis Public Schools
Indianapolis, Indiana

For information regarding the CPSIA on this printed material
call 203-595-3636 and provide reference # RICH-305765-4T

ISBN-13: 978-1-888287-70-7
Copyright © 2008, © 2009 by Calvert School, Inc.
Printed in China
1 2 3 4 5 6 7 8 9 10   12 11 10 09 08

To the Student

This book was made for you. It will teach you many new things about mathematics. Your math book is a tool to help you discover and master the skills you will need to become powerful in math.

About the Art in this Book

Calvert home schooling students from all over the world and Calvert Day School students have contributed their original art for this book. We hope you enjoy looking at their drawings as you study mathematics.

CONTENTS

Diagnosing Readiness

1 Numbers and Place Value

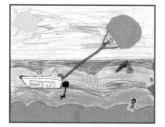

2 Adding and Subtracting Whole Numbers

3 Multiplying by Whole Numbers

4 Multiplying by Two-Digit Numbers

5 Statistics and Data

6 Dividing by One-Digit Numbers

7 Dividing by Two-Digit Numbers

8 Measurement

9 Geometry

10 Fractions and Probability

11 Adding and Subtracting Fractions

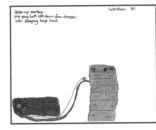

12 Adding and Subtracting Decimals

Diagnosing Readiness

In this preliminary chapter, you will take a diagnostic skills pretest, practice skills that you need to work on, and take a diagnostic skills posttest. This preliminary chapter will assure you have the skills necessary to begin Chapter 1.

Diagnostic Skills Pretest

Complete the following pretest. If you miss one or more problems from any section, refer to that section in the diagnostic chapter.

Write each number in standard form, expanded form, and words. (Section 1)

1. two thousand

2. seven thousand, four hundred twenty-nine

3. 5,000 + 400 + 2

4. 6,372

5. 8,005

Round each number to the underlined place-value position. (Section 2)

6. <u>3</u>42

7. 5,9<u>2</u>7

8. <u>9</u>,037

9. 5,<u>3</u>80

Count each amount of money. Then write the smaller amount. (Section 3)

10. a.

b.

Tell the time shown by each clock. (Section 4)

11.

12.

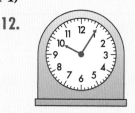

Write the addition or subtraction sentence described, then solve. (Section 5)

13. Carol bought 5 books. Sam bought 19 books. How many more books did Sam buy?

14. Sara baked 12 blueberry muffins and 9 apple muffins. How many muffins did she bake in all?

Write a related fact. (Section 6)

15. $8 + 6 = 14$

16. $9 - 4 = 5$

Write the fact family. (Section 6)

17. 3, 6, 9

18. 12, 8, 4

Add or subtract. (Section 7–8)

19. $\begin{array}{r} 3,720 \\ + 1,274 \\ \hline \end{array}$

20. $\begin{array}{r} 7,926 \\ + 162 \\ \hline \end{array}$

21. $\begin{array}{r} 4,872 \\ - 3,551 \\ \hline \end{array}$

22. $\begin{array}{r} 2,385 \\ - 1,472 \\ \hline \end{array}$

Multiply. (Section 9)

23. 6×7

24. 3×4

25. 0×9

26. 8×1

State how many are in each group or how many groups there are. (Section 10)

27. There are 15 people on a city tour. There are 3 groups. How many are in each group?

28. There are 20 people rafting down the river. There are 5 people in each raft. How many rafts are there?

Divide. (Section 10 and 11)

29. $25 \div 5$

30. $28 \div 4$

31. $9 \div 9$

32. $0 \div 6$

Use grid paper to show each decimal or fraction. (Section 12)

33. 0.4

34. $\frac{7}{10}$

35. 0.1

Use the calendar to answer the questions. (Section 13)

36. On what day of the week is August 12?

37. If Caleb is having a party two weeks from August 3, on what date is the party?

August

S	M	T	W	T	F	S
				1	2	3
4	5	6	7	8	9	10
11	12	13	14	15	16	17
18	19	20	21	22	23	24
25	26	27	28	29	30	31

Favorite Bicycle Color

Use the bar graph to answer questions 40–41. (Section 14)

38. How many children chose pink as their favorite bicycle color?

39. What color received the least number of votes?

Diagnostic Skills Pretest xv

1 Place Value

The town of Oakley hosted its annual summer fair. There were 2,346 people who attended the fair.

We use digits and place value to write numbers.

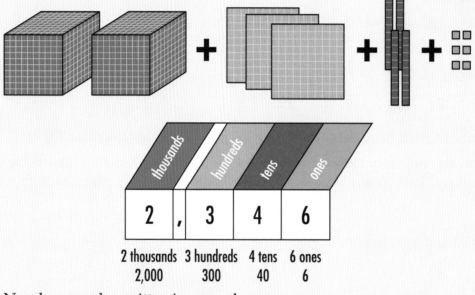

	thousands		hundreds	tens	ones
	2	,	3	4	6

2 thousands 3 hundreds 4 tens 6 ones
2,000   300   40   6

Numbers can be written in several ways.

Standard Form ⟶ 2,346
Expanded Form ⟶ 2,000 + 300 + 40 + 6
Words ⟶ two thousand, three hundred forty-six

TRY THESE

Write each number in standard form.

1.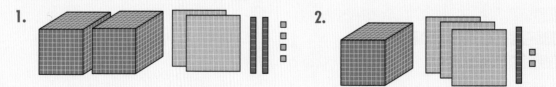

2.

3.

4.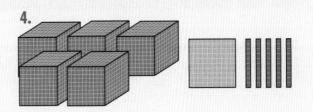

Exercises

Write each number in standard form and expanded form.

1. five thousand

2. three thousand, two hundred

3. nine hundred eighty-one

4. fifty-eight

5. one thousand, six hundred forty

6. two thousand, fifty

Write each expanded form in standard form and words.

7. 3,000 + 40 + 8

8. 6,000 + 200 + 40 + 1

9. 5,000 + 4

10. 2,000 + 10 + 3

Write each number in expanded form and words.

11. 665

12. 7,520

13. 5,205

14. 9,008

Write the value named by each 4.

15. 425

16. 3,942

17. 2,804

18. 6,425

PROBLEM SOLVING

19. Cal Ripken played 2,632 consecutive games for the Baltimore Orioles. Write this number in expanded form and words.

20. Draw a picture to show 3 thousands, 7 hundreds, 4 tens, and 9 ones. Write the number in expanded form and standard form.

2 Rounding

Mr. Cedeno's class went on a field trip to the state capitol building. There are 28 students in his class. Mr. Cedeno told the guide he has about 30 students. Mr. Cedeno rounded 28 to the nearest 10.

You can use a number line to explore rounded numbers.

> Rounded numbers tell *about* how many.

```
←——+——+——+——+——+——+——+——+——+——+——→
   20  21  22  23  24  25  26  27  28  29  30
```

You can see on the number line that 28 rounds to 30 when it is rounded to the nearest 10. 28 is closer to 30 than to 20.

You can round numbers to the nearest 10, 100, 1,000, and so on.

To round a number, circle the number in the position to which you are rounding.

Then look at the digit one place to the right of the circled digit.

If that number is less than 5, you round down.

If that number is 5 or more, you round up.

Examples

A. Round 3,600 to the nearest thousand.

③,600   Circle the 3.

3,600   Look at the digit to the right of the 3.

          It is 6. 6 is greater than 5.

3,600 rounds up to 4,000.

B. Round 819 to the nearest hundred.

⑧19   Circle the 8.

819   Look at the digit to the right of the 8.

          It is 1. 1 is less than 5.

819 rounds down to 800.

TRY THESE

Use the number line to round each number to the nearest hundred.

400   450   500

1. 430   **2.** 450   **3.** 409   **4.** 482   **5.** 472

Exercises

Round each number to the tens place.

1. 73   **2.** 529   **3.** 51   **4.** 7,548

Round each number to the underlined place-value position.

1. 4<u>7</u>2   **2.** <u>8</u>17   **3.** <u>6</u>2   **4.** <u>1</u>,820

5. 7,02<u>7</u>   **6.** 2,<u>6</u>00   **7.** 8,6<u>5</u>2   **8.** <u>3</u>6

9. 3<u>0</u>3   **10.** 5,<u>2</u>60   **11.** <u>2</u>80   **12.** 2,<u>4</u>91

PROBLEM SOLVING

13. Round the lengths to the nearest thousand.

14. Round the lengths to the nearest hundred.

15. Round the lengths to the nearest ten.

U.S. Bridges	
Golden Gate Bridge	8,981 feet long
Francis Scott Key Bridge	9,036 feet long
New River Gorge	3,031 feet long
Bayonne	8,460 feet long
Fremont	8,063 feet long

3 Count, Compare, and Order Money

Caleb wants to buy a toy train for his train collection. The train costs $4.55. He has 4 one-dollar bills, 1 quarter, and 4 dimes. Does he have enough money to buy the train?

Count the money below to find out if Caleb has enough to buy the train.

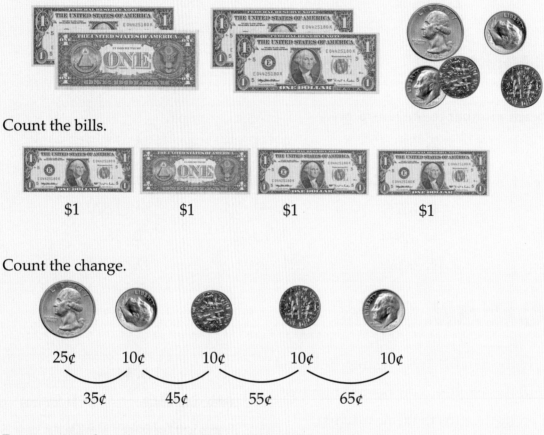

Count the bills.

$1 $1 $1 $1

Count the change.

25¢   10¢   10¢   10¢   10¢

      35¢   45¢   55¢   65¢

By counting the money, you can see that Caleb has $4.65.

$4.65 > $4.55   Compare the amounts.

Caleb can buy the train.

Another Example

Ben, Carla, and Mark have the following amounts of money in their piggy banks. Put the amounts in order from least to greatest.

Ben **Carla** **Mark**

When you count the coins in each bank you see that Ben has 87¢, Carla has 86¢, and Mark has 73¢.

In order from least to greatest the values are 73¢, 86¢, and 87¢.

TRY THESE

Count the money then write the amount.

1.

2.

Exercises

Count and write each amount of money. Then write the greater amount.

1. a.

   b.

2. a.

   b.

3. a.

   b.

4. a.

   b.

Write the total amount of money in each piggy bank. Put the amounts of money in order from least to greatest.

5. **Bank A** 4 $1 bills, 4 nickels

   Bank B 4 $1 bills, 2 dimes, 2 pennies

   Bank C 4 $1 bills, 1 dime and 3 nickels

PROBLEM SOLVING

Use the information in the chart to answer questions 6–8.

6. List the prices of Food and Drink items in order from least to greatest.

7. Dana has 4 $1 bills, 3 quarters, and two dimes. Does she have enough money to buy a child's ticket?

8. Peter has 3 $1 bills. Does he have enough money to buy candy and popcorn?

Big Top Circus

Tickets	
Adult	$7.50
Child	$5.25
Food and Drink	
Juice	$1.25
Candy	$0.75
Popcorn	$2.10

4 Time to Five Minutes

Paul arrived at school at 8:05 in the morning. This time can be represented on a clock. Look at the clock to the right.

When you look at a clock you can tell time to the nearest five minutes before or after the hour. Look at the clocks below.

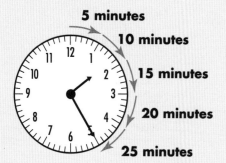

You can count forward. Each number on the face of the clock represents 5 minutes.

This clock shows 1:25.

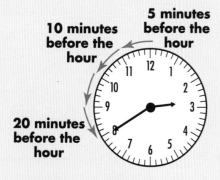

You can also tell time before the next hour.

This clock shows 2:40 or 20 minutes before 3.

TRY THESE

Tell the time shown by each clock.

1.

2.

Exercises

Tell the time shown by each clock.

1.

2.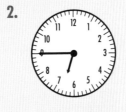

3.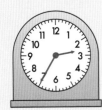

4.

PROBLEM SOLVING

5. Daniel ate lunch at 12:25. Draw a clock and show 12:25 on the clock.

6. Kaitlin asked her mom to pick her up from the mall at 4:50. Kaitlin's mom arrived at ten minutes before five. Did Kaitlin's mom arrive at the right time?

7. Jacob's karate class starts at 6:55. Draw a clock and show 6:55 on the clock.

8. A clock shows 9:50. How would you find the number of minutes left in the hour?

5 Addition and Subtraction

Sandy enjoys music. She has 4 CDs. She gets 2 *more* for her birthday. *How many* CDs does she have *altogether*?

There are 10 CDs on the shelf. Sandy takes 3 CDs off the shelf. How many CDs are *left* on the shelf?

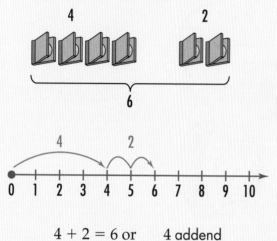

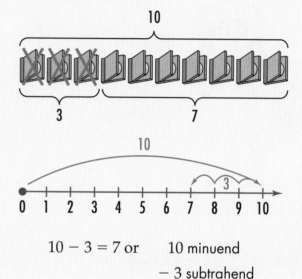

$4 + 2 = 6$ or

$$\begin{array}{r} 4 \text{ addend} \\ + 2 \text{ addend} \\ \hline 6 \text{ sum} \end{array}$$

$10 - 3 = 7$ or

$$\begin{array}{r} 10 \text{ minuend} \\ - 3 \text{ subtrahend} \\ \hline 7 \text{ difference} \end{array}$$

More Examples

A. After a sale there were 7 rock albums left unsold and 6 jazz albums left unsold. How many albums were not sold?

Addition $7 + 6 = 13$

13 albums were not sold.

B. Carlos had 8 balloons. He gave 3 away to Lucy. How many balloons did Carlos keep?

Subtraction $8 - 3 = 5$

Carlos kept 5 balloons.

TRY THESE

Solve.

1. There are 16 horses in the pasture. Rod puts 9 horses in the barn. How many horses are left in the pasture?

2. Bill has 10 stamps in his collection. He buys 3 more. What is the total number of stamps in his collection now?

Exercises

Write the addition or subtraction sentence that is shown in each diagram. Then solve.

1.
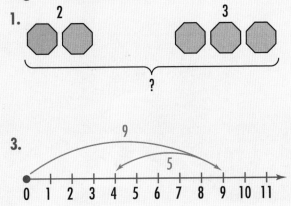

2.
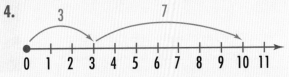

3.

4.
3   7

0 1 2 3 4 5 6 7 8 9 10 11

5. Brenda has 17 stamps. Fred has 8 stamps. How many more stamps does Brenda have than Fred?

6. Shawn has 6 goldfish and 8 guppies. How many fish does Shawn have?

7. Robin had 15 butterflies in her collection. She gave her sister 6 butterflies. How many butterflies does Robin have left?

8. Molly has 6 baseball cards. Janet gives her 3 more. How many baseball cards does Molly have altogether?

PROBLEM SOLVING

Choose the number sentence that solves each problem. Then answer each question.

9. There are 14 plants and 8 pots. Each plant needs a pot. How many more pots are needed?

   a. $14 + 8 = 22$ b. $14 - 8 = 6$ c. $14 - 6 = 8$

10. After dinner there were 4 apples and 3 oranges left. How many pieces of fruit were left?

    a. $4 + 3 = 7$ b. $7 - 4 = 3$ c. $7 - 3 = 4$

6 Fact Families

James caught 11 fish. He caught 5 catfish. All the others were sunfish. How many were sunfish?

Addition and subtraction are related. Since $5 + 6 = 11$, then $11 - 5 = 6$. James caught 6 sunfish.

$$5 + 6 = 11 \qquad 11 - 5 = 6$$
and $\quad 6 + 5 = 11 \qquad 11 - 6 = 5$

Number facts that use the same numbers are called a *fact family*.

More Examples

A. $7 + 4 = 11$

$4 + 7 = 11$

$11 - 7 = 4$

$11 - 4 = 7$

B. $5 + 0 = 5$

$0 + 5 = 5$

$5 - 0 = 5$

$5 - 5 = 0$

C. $6 + 6 = 12$

$12 - 6 = 6$

TRY THESE

Match the related facts.

1. $2 + 0 = 2$ **a.** $8 - 4 = 4$

2. $4 + 4 = 8$ **b.** $8 + 7 = 15$

3. $9 - 3 = 6$ **c.** $2 - 0 = 2$

4. $15 - 7 = 8$ **d.** $1 - 0 = 1$

5. $1 + 0 = 1$ **e.** $3 + 6 = 9$

6. $7 + 6 = 13$ **f.** $13 - 7 = 6$

Exercises

Write a related fact.

1. $7 + 0 = 7$ 2. $14 - 7 = 7$ 3. $3 + 2 = 5$

4. $9 + 4 = 13$ 5. $18 - 9 = 9$ 6. $7 + 3 = 10$

7. $0 + 1 = 1$ 8. $2 + 6 = 8$ ★9. $8 - 0 = 8$

Write the fact family.

10. 4, 6, 10   11. 0, 3, 3   12. 5, 3, 8

13. 17, 9, 8   14. 3, 8, 11   15. 7, 6, 1

16. 4, 3, 7   17. 8, 4, 4   18. 13, 9, 4

19. Peter has 7 fish in his aquarium. His mother gives him 7 more fish. How many fish does he have in the aquarium altogether?

20. Dee planted 16 watermelon seeds. Only 7 plants grew to be mature plants. How many seeds did not grow?

21. Barbara earns $5 selling lemonade. She spends $3 at the arcade. How much money does she have left?

22. At the airport, 11 planes take off in an hour. Eight planes land that same hour. How many more planes took off than landed?

7 Adding 3–4 Digit Numbers With or Without Regrouping

Jill's favorite display at the museum has 1,672 butterflies and 2,589 moths. What is the total number of moths and butterflies in the display?

To find the total, add 1,672 and 2,589.

THINK
An estimate is
$2,000 + 3,000 = 5,000.$

Step 1	Step 2	Step 3
Add the ones and tens.	Add the hundreds.	Add the thousands.
$\begin{array}{r} {}^{1}{}^{1}\\ 1,672 \\ +\ 2,589 \\ \hline 61 \end{array}$	$\begin{array}{r} {}^{1}{}^{1}{}^{1}\\ 1,672 \\ +\ 2,589 \\ \hline 261 \end{array}$	$\begin{array}{r} {}^{1}{}^{1}{}^{1}\\ 1,672 \\ +\ 2,589 \\ \hline 4,261 \end{array}$

The display has 4,261 moths and butterflies.   Is the answer reasonable?

More Examples

Estimate to see if the sum is reasonable.

A.
$\begin{array}{r} {}^{1\,1}\\ 8,279 \\ +\ 6,537 \\ \hline 14,816 \end{array}$

$8,000 + 7,000 = 15,000$

B.
$\begin{array}{r} {}^{1\,1}\\ 1,146 \\ +\ 8,285 \\ \hline 9,431 \end{array}$

$1,000 + 8,000 = 9,000$

C.
$\begin{array}{r} {}^{1\ \ 1}\\ \$6,977 \\ +\ 2,814 \\ \hline \$9,791 \end{array}$

$\$7,000 + \$3,000 = \$10,000$

TRY THESE

Add.

1. $\begin{array}{r} 587 \\ +\ 503 \end{array}$

2. $\begin{array}{r} \$8.76 \\ +\ 6.86 \end{array}$

3. $\begin{array}{r} 2,787 \\ +\ 3,615 \end{array}$

4. $\begin{array}{r} \$4,489 \\ +\ 8,37 \end{array}$

5. $\begin{array}{r} 2,948 \\ +\ 295 \end{array}$

6. $\begin{array}{r} 7,502 \\ +\ 386 \end{array}$

7. $\begin{array}{r} \$8,479 \\ +\ 776 \end{array}$

8. $\begin{array}{r} 736 \\ +\ 6,849 \end{array}$

9. $\begin{array}{r} 2,846 \\ +\ 478 \end{array}$

10. $\begin{array}{r} 5,794 \\ +\ 842 \end{array}$

Exercises

Add.

1.	2,474	2.	2,275	3.	4,685	4.	$723	5.	2,454
	+ 5,487		+ 6,815		+ 2,543		+ 168		+ 5,489

6.	8,428	7.	$482	8.	4,824	9.	7,089	10.	1,385
	+ 786		+ 531		+ 1,343		+ 4,502		+ 6,757

11. 2,831 + 1,543

12. 6,821 + 3,025

PROBLEM SOLVING

13. A science lab has 458 cocoons. It has 576 more caterpillars than cocoons. How many caterpillars are in the lab?

14. The museum gift shop sold 2,178 postcards in June. In July, 3,492 postcards were sold. How many postcards were sold in all?

8 Subtracting 2+ Digits With and Without Regrouping

Jody Steele's parents are beekeepers. She helps them take care of their 55 beehives. They would like to have 170 beehives. How many more must they buy? Subtract 55 from 170.

▶ The technical name for a beekeeper is *apiarist*.

THINK An estimate is 170 − 60 = 110.

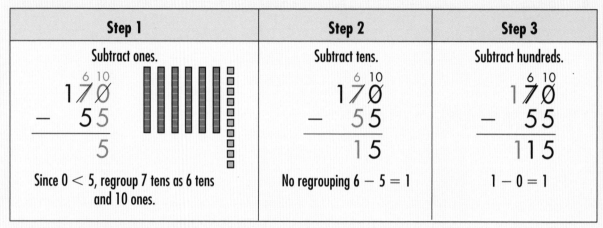

Step 1	Step 2	Step 3
Subtract ones.	Subtract tens.	Subtract hundreds.

6 10
17̸0̸
− 55
———
        5
	6 10
	17̸0̸
	− 55
	———
	      15
		6 10
		17̸0̸
		− 55
		———
		    115

Since 0 < 5, regroup 7 tens as 6 tens and 10 ones.	No regrouping 6 − 5 = 1	1 − 0 = 1

They must buy 115 beehives.   Is the answer reasonable?

More Examples

A.
```
  3 9 9 13
 $4,003
− 1,374
────────
 $2,629
```
Check your answer with the *inverse* operation, addition.

```
 1 11
 2,629
+ 1,374
───────
 4,003
```

B.
```
 $3,798
− 1,523
────────
 $2,275
```

C.
```
      10
   7 0 16
  2,8̸1̸6̸
−   7̸7̸8
────────
  2,038
```

TRY THESE

. . . .

Write the difference.

1.
```
  514
  6̸4̸
− 28
```

2.
```
  413
  5̸3̸
−  8
```

3.
```
     17
  3 7̸16
 $4̸8̸6̸
−  97
```

4.
```
  910
  1̸0̸0̸
−  67
```

5.
```
  3 917
 9,4̸0̸7̸
−    89
```

6.
```
 477
−  9
```

7.
```
  86
− 47
```

8.
```
 $58
− 29
```

9.
```
  92
− 35
```

10.
```
 $475
− 182
```

Exercises

Subtract.

1. 836
   − 27

2. $741
   − 203

3. 6,154
   − 36

4. 1,384
   − 265

5. 7,656
   − 2,348

6. $340
   − 196

7. 1,652
   − 837

8. 900
   − 45

9. 7,431
   − 358

10. 6,665
    − 2,758

11. 8,364
    − 4,527

12. 3,895
    − 1,672

13. $4,450
    − 3,295

14. 4,000
    − 52

15. $9,004
    − 5,478

16. $5,978 - \blacksquare = 1,394$

17. $\$5,274 - \blacksquare = \$3,781$

18. $\blacksquare - 1,395 = 4,605$

19. $1,700 - (234 + 678)$

20. $(1,700 - 234) + 678$

21. Find the difference of 56,374 and 4,932.

22. Without subtracting, write which is greater, $800 - 542$ or $800 - 245$. Explain why you know your answer is correct.

PROBLEM SOLVING

23. What is the total number of people who live in both Lewiston and Mayville?

24. How many more people live in Lewiston than in Mayville?

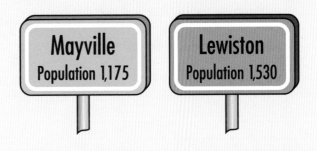

Mayville
Population 1,175

Lewiston
Population 1,530

9 Basic Multiplication Facts

How many stamps are shown below?

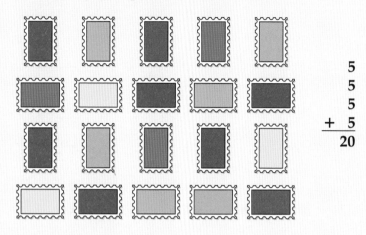

$$\begin{array}{r} 5 \\ 5 \\ 5 \\ + \ 5 \\ \hline 20 \end{array}$$

$4 \times 5 =$ ■ $4 \times 5 = 20$ There are 20 stamps.

$\underbrace{4 \times 5}_{\text{factors}} = \underbrace{20}_{\text{product}}$

A **factor** is any number that you multiply. The factors are 4 and 5 in the first example.

A **product** is the answer to a multiplication problem. For example, there are 20 stamps. $4 \times 5 = 20$

More Examples

A. 2 rows of 8

$2 \times 8 = 16$

B. 6 groups of 7

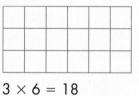

$6 \times 7 = 42$

C. a 3-by-6 rectangle

$3 \times 6 = 18$

TRY THESE

Complete each pattern.

1. 3, 6, 9, ■, ■

2. 5, 10, 15, ■, ■

3. 4, 8, 12, ■, ■

4. 18, 24, 30, ■, ■

5. 21, 28, 35, ■, ■

6. 27, 36, 45, ■, ■

Write the multiplication fact shown by each picture.

7.

8.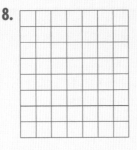

9. X X X X X X X X X
X X X X X X X X X
X X X X X X X X X
X X X X X X X X X

Exercises

Multiply.

1.	3 × 2	2.	4 × 7	3.	$9 × 2	4.	3 × 3	5.	0 × 8	6.	7 × 4

7.	9 × 1	8.	$6 × 4	9.	3 × 8	10.	5 × 3	11.	$9 × 5	12.	9 × 2

13. 5×2

14. 7×7

15. 5×8

16. 9×8

17. $2 \times 2 \times 3$

18. $2 \times 4 \times 2$

19. $3 \times 3 \times 6$

20. In $2 \times 9 = 18$, what are the numbers 2 and 9 called?

21. The factors are 5 and 4. What is the product?

22. List all the factors of 8.

23. List all the factors of 25.

24. List all the factors of 36.

PROBLEM SOLVING

25. Jay put flowers in 5 vases. He put 7 flowers in each vase. How many flowers did he use?

26. Rhett has 9 airmail stamps and 4 commemorative stamps. How many more airmail stamps than commemorative stamps does he have?

27. Della can read 3 pages in a minute. How many pages can she read in 8 minutes?

28. Luis works 8 hours a day. How many hours does he work in 5 days?

10 Division

Twenty-four children are in line to ride the school bus. Three children will sit in each seat. How many seats are needed?

To answer this question, you need to find out how many groups of three are formed. You can use counters or draw a picture to see how many groups of three can be made from twenty-four. The X's below represent the twenty-four children.

XXX XXX XXX XXX XXX XXX XXX XXX

You can see that there are eight groups of three. This means that twenty-four divided by three is eight. This can also be represented by the following number sentence, $24 \div 3 = 8$.

In some cases you want to know how many are in a group. If you have 8 people and you want to divide them into 4 equal groups, you need to find out how many are in each group.

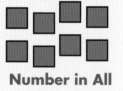

Number in All
8 people

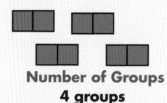

Number of Groups
4 groups

Number in Each Group
2 people in each group

This means that $8 \div 4 = 2$.

TRY THESE

Complete the table below.

	Number in All	Number of Groups	Number in Each Group
1.	42 people	7 cars	■ people in each car
2.	15 airplanes	5 hangars	■ airplanes in each hangar
3.	35 marbles	5 players	■ marbles for each player
4.	20 minutes	2 speeches	■ minutes for each speech
5.	30 dollars	5 baseballs	■ dollars for each baseball

Show the "Number in All" with base-ten blocks or counters.
Then separate your blocks or counters into the "Number in Each
Group." Record your "Number of Groups."

	Number in All	Number of Groups	Number in Each Group
1.	36 students	■ seats	3 students in each seat
2.	10 rabbits	■ cages	2 rabbits in each cage
3.	8 buttons	■ collars	1 button on each collar
4.	28 tickets	■ children	7 tickets for each child
5.	66 tires	■ buses	6 tires on each bus

PROBLEM SOLVING

6. There are 8 pages in a photo album and 40 photos. If you place an equal number of photos on each page, how many photos can you place on each page?

7. Discuss the steps you took to find the Number in Each Group in **Try These**.

8. Discuss the steps you took to find the Number of Groups in problems 1– 5.

9. If a classroom has 24 children sitting in 4 rows, how many children are in each row?

10. Use the numbers 30, 5, and 6 to make up two "number in each group" problems. Use base-ten blocks, counters, or drawings to show your problems.

11 Basic Division Facts

On a recent delivery, Mr. Armas drove 42 miles. His truck used 6 gallons of gasoline. How many miles did Mr. Armas drive for each gallon?

There are two ways to write a division sentence.

$$42 \div 6 = \blacksquare \qquad \text{gallons} \longrightarrow 6\overline{)42} \longleftarrow \text{miles}$$

Each part of the division sentence has a special name.

$$\underset{\text{dividend}}{42} \div \underset{\text{divisor}}{6} = \underset{\text{quotient}}{7}$$

$$6\overline{)42} \quad \begin{array}{l} \text{quotient} \\ \text{dividend} \end{array}$$

divisor

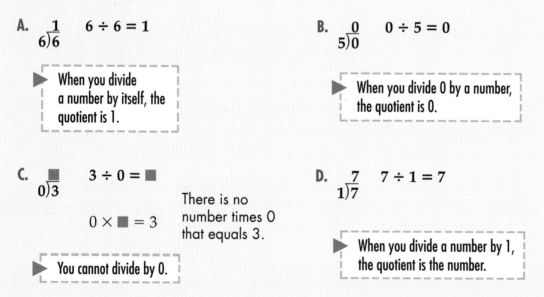

Mr. Armas drove 7 miles for each gallon used, or 7 miles per gallon.

There are special rules when you divide with 1 and 0.

More Examples

A. $\dfrac{1}{6\overline{)6}}$ $6 \div 6 = 1$

> When you divide a number by itself, the quotient is 1.

B. $\dfrac{0}{5\overline{)0}}$ $0 \div 5 = 0$

> When you divide 0 by a number, the quotient is 0.

C. $\dfrac{\blacksquare}{0\overline{)3}}$ $3 \div 0 = \blacksquare$

$0 \times \blacksquare = 3$ There is no number times 0 that equals 3.

> You cannot divide by 0.

D. $\dfrac{7}{1\overline{)7}}$ $7 \div 1 = 7$

> When you divide a number by 1, the quotient is the number.

TRY THESE

Find the quotient.

1. $12 \div 6$
2. $14 \div 7$
3. $64 \div 8$
4. $63 \div 9$
5. $28 \div 7$

6. $9\overline{)0}$
7. $8\overline{)72}$
8. $9\overline{)81}$
9. $8\overline{)8}$
10. $6\overline{)36}$

Exercises

Divide.

1. $0 \div 8$
2. $35 \div 7$
3. $18 \div 9$
4. $16 \div 4$
5. $4 \div 1$

6. $6 \div 2$
7. $7 \div 7$
8. $32 \div 8$
9. $30 \div 5$
10. $27 \div 9$

11. $7\overline{)0}$
12. $3\overline{)3}$
13. $5\overline{)45}$
14. $6\overline{)54}$
15. $3\overline{)21}$

16. $6\overline{)48}$
17. $7\overline{)56}$
18. $9\overline{)36}$
19. $9\overline{)0}$
20. $3\overline{)18}$

21. $7\overline{)63}$
22. $5\overline{)40}$
23. $7\overline{)42}$
24. $9\overline{)72}$
25. $2\overline{)14}$

26. $4\overline{)24}$
27. $1\overline{)2}$
28. $8\overline{)64}$
29. $4\overline{)28}$
30. $8\overline{)24}$

31. What is 81 divided by 9?
32. Find $32 \div 4$.

PROBLEM SOLVING

33. Mr. Harvey has 36 plants. He wants to place them on 9 shelves, with the same number on each shelf. How many plants will Mr. Harvey place on each shelf?

34. Jenny had 16 chocolate bars. After she gave 4 bars to each of her brothers, she had none left. How many brothers did Jenny have?

35. Jon spent 45¢ for pencils. How many pencils did he buy?

36. There are 28 dancers and 7 singers in a musical group. How many more dancers than singers are in the group?

An American inventor, Clarence Birdseye, developed the quick-freezing process of preserving food in the 1920s. In three out of Ms. Hu's last ten meals, she used frozen foods. You can write a fraction to show what part of her meals were from frozen foods.

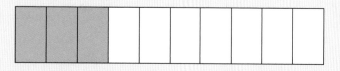

numerator → **3** ← number you are talking about
denominator → **10** ← total number

Three tenths of the meals were from frozen foods.

You can write $\frac{3}{10}$ as a **decimal.**

A fraction that has a denominator of 10, 100, and so on can be written as a decimal.

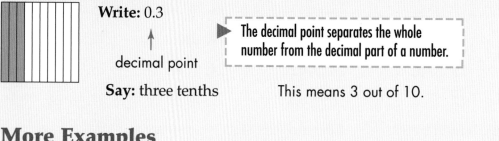

Write: 0.3

↑
decimal point

Say: three tenths

▶ The decimal point separates the whole number from the decimal part of a number.

This means 3 out of 10.

More Examples

A. $\frac{4}{10}$ or 0.4

B. A fraction shows division. For example, the fraction $\frac{8}{10}$ is another way of saying $8 \div 10$.

$\frac{8}{10}$ or 0.8

TRY THESE

Write a fraction and a decimal to name each shaded part.

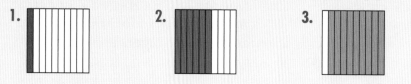

1.   **2.**   **3.**   **4.**

Exercises

Write a fraction and a decimal to name each shaded part.

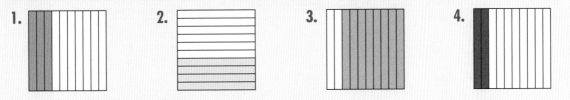

1.   2.   3.   4.

Use grid paper to show each decimal or fraction.

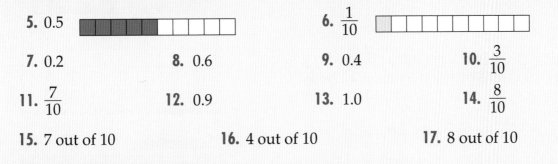

5. 0.5   6. $\frac{1}{10}$

7. 0.2   8. 0.6   9. 0.4   10. $\frac{3}{10}$

11. $\frac{7}{10}$ 12. 0.9   13. 1.0   14. $\frac{8}{10}$

15. 7 out of 10   16. 4 out of 10   17. 8 out of 10

PROBLEM SOLVING

18. On Monday, at the vet, 7 out of 10 dogs got a vaccination. Write a fraction to show what part of the group of dogs got a vaccination.

19. At a store, 9 out of 10 pizzas that are sold are frozen pizzas. Write a fraction to show what part of the pizzas sold are frozen pizzas.

13 Using a Calendar

Margaret is looking on the calendar to on see what day of the week her birthday falls this year. Her birthday is on July 5th. On what day of the week does Margaret's birthday fall this year?

July

S	M	T	W	T	F	S
	1	2	3	Independence Day 4	5	6
7	8	9	10	11	12	13
14	15	16	17	18	19	20
21	22	23	24	25	26	27
28	29	30	31			

A calendar shows the days of the week and the months in a year. You can use a calendar to see on what day of the week a special event happens. You can also use a calendar to plan your schedule, to plan vacations, or see when holidays will take place.

By looking at the July calendar to the right, you can see that Margaret's birthday falls on a Friday.

August

S	M	T	W	T	F	S
				1	2	3
4	5	6	7	8	9	10
11	12	13	14	15	16	17
18	19	20	21	22	23	24
25	26	27	28	29	30	31

You should know that the months in order are

January	**February**	**March**	**April**
May	**June**	**July**	**August**
September	**October**	**November**	**December**

September

S	M	T	W	T	F	S
1	Labor Day 2	3	4	5	6	7
8	9	10	11	12	13	14
15	16	17	18	19	20	21
22	23	24	25	26	27	28
29	30					

Another Example

Margaret's family is taking a 2-week vacation in August. If they begin their vacation on August 14th, when will they return?

You can look at the calendar and count 2 weeks from August 14th. Their vacation will end on August 28th.

TRY THESE

Find each date. Use the calendars above.

1. 1 week after July 12th

2. 8 days after September 15th

3. 12 days after August 2nd

4. 14 days before July 30th

Exercises

Use the calendars on the previous page to solve.

1. Graham leaves on July 1st to go to summer camp for 3 weeks. When does he return from summer camp?

2. Sophie is having her birthday party 5 days after her birthday. If her party is on July 27th, when is her birthday?

3. How many days pass from the third Tuesday in August to the second Tuesday in September?

4. School starts on the first Tuesday after Labor Day. When does school start?

5. The Carsons drove to Wyoming. They arrived on July 22nd. If the drive took 4 days, when did they leave for Wyoming?

14 Reading Bar Graphs

A bar graph is one way to compare information.

Using this bar graph, you can compare the heights of four dams.

Use the graph to find the height of each dam. Which is higher, Hoover Dam or Oroville Dam?

| 1. READ | You need to know which dam is higher. Read the title of the bar graph. It compares the heights of dams. |

| 2. PLAN | Read the horizontal scale to find the two dams you are comparing. Read the vertical scale. |

| 3. SOLVE | Compare the bars on the graph. The bar for Oroville Dam is higher than the bar for Hoover Dam. |

| 4. CHECK | Use the graph to find the height of each dam. Hoover–730 feet   Oroville–770 feet
Compare the heights. Oroville Dam is higher. 730 < 770 |

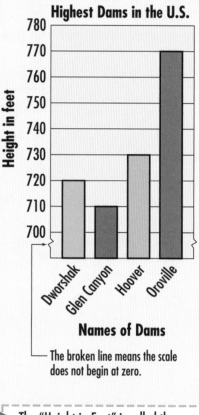

Highest Dams in the U.S.

The broken line means the scale does not begin at zero.

▶ The "Height in Feet" is called the *vertical scale* and "Names of Dams" is called the *horizontal scale*.

TRY THESE

1. Which dam is higher, Glen Canyon or Hoover?

2. Which dam is the third highest in the United States?

3. List the names of the dams from highest to lowest.

4. Is the Oroville Dam twice as tall as the Hoover Dam? Explain your answer.

Solve..

Use the bar graphs to answer the questions.

Distance to the Library (in blocks)

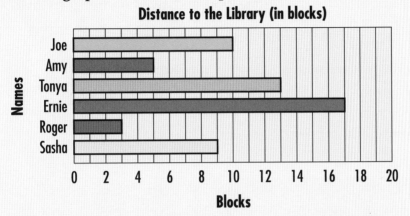

1. Which student lives farthest from the library?

2. Which student lives nearest to the library?

3. Which students live less than 8 blocks from the library?

4. Which students live more than 8 blocks from the library?

5. What is this bar graph measuring?

6. Which building has the fewest stories?

7. Which building has exactly 100 stories?

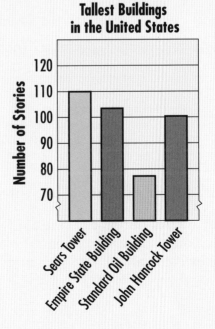

Diagnostic Skills Posttest

Complete the problems below.

Write each number in standard form, expanded form, and words. (Section 1)

1. three thousand, four hundred

2. nine thousand, two hundred forty-five

3. 3,000 + 900 + 22

4. 4,803

5. 6,010

Round each number to the underlined place-value position. (Section 2)

6. 8<u>2</u>0

7. 4,<u>6</u>93

8. <u>6</u>,825

9. 5,0<u>3</u>9

Count and write each amount of money. Then write the smaller amount. (Section 3)

10. a.

b.

Tell the time shown by each clock. (Section 4)

11.

12.

Write the addition or subtraction sentence described, then solve. (Section 5)

13. David ate 6 donut holes. Jana ate 8 donut holes. How many donut holes did they eat in all?

14. Wendy picked 15 roses from the garden. Kim picked 8 roses from the garden. How many more roses did Wendy pick than Kim?

Write a related fact. (Section 6)

15. $16 - 9 = 7$

16. $3 + 8 = 11$

Write the fact family. (Section 6)

17. $5, 5, 10$

18. $17, 8, 9$

Add or subtract. (Section 7–8)

19.
$$\begin{array}{r} 2{,}739 \\ +\ 4{,}001 \\ \hline \end{array}$$

20.
$$\begin{array}{r} 4{,}829 \\ +\ 523 \\ \hline \end{array}$$

21.
$$\begin{array}{r} 8{,}697 \\ -\ 2{,}354 \\ \hline \end{array}$$

22.
$$\begin{array}{r} 5{,}421 \\ -\ 2{,}322 \\ \hline \end{array}$$

Multiply. (Section 9)

23. 5×8

24. 9×7

25. 3×1

26. 6×2

State how many are in each group or how many groups there are. (Section 10)

27. There are 32 cars parked on a lot in 8 rows. How many cars are in each row?

28. You have \$25. Each soccer ball cost \$5. How many soccer balls can you buy?

Divide. (Section 11)

29. $48 \div 6$

30. $35 \div 7$

31. $0 \div 4$

32. $10 \div 2$

Use grid paper to show each decimal or fraction. (Section 12)

33. 0.6

34. $\dfrac{4}{10}$

35. 0.2

Find each date. Use the calendar to the right to answer questions 36–37. (Section 13)

36. 3 weeks after January 4th

37. 10 days before January 19th

January

S	M	T	W	T	F	S
	1	2	3	4	5	6
7	8	9	10	11	12	13
14	15	16	17	18	19	20
21	22	23	24	25	26	27
28	29	30	31			

Use the bar graph below to answer questions 38–39. (Section 14)

38. How many children voted for sugar cookies?

39. What kind of cookie did most children vote for?

Favorite Cookies

Kind of Cookie / Number of Votes

Numbers and Place Value

Matthew Klas
Minnesota

1.1 Numbers Through Millions

Objective: to apply knowledge of place value to the millions period

Chicago, Illinois, is one of the largest cities in the United States. The population of Chicago is approximately 6,945,872 people.

You use digits and **place value** to write numbers.

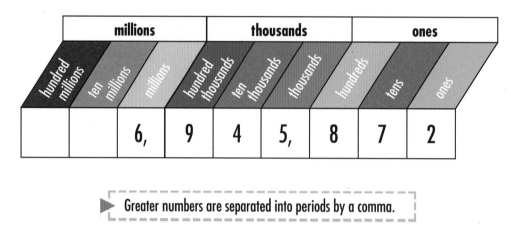

> ▶ Greater numbers are separated into periods by a comma.

Each group of three digits, starting from the ones place, is called a **period**. When writing numbers, the periods are separated by commas.

You can write numbers in three ways:

Standard Form → 6,945,872

Expanded Form → 6,000,000 + 900,000 + 40,000 + 5,000 + 800 + 70 + 2

Words ⟶ six million, nine hundred forty-five thousand, eight hundred seventy-two

Another Example

Standard Form → 9,231,002

Expanded Form → 9,000,000 + 200,000 + 30,000 + 1,000 + 2

Words ⟶ nine million, two hundred thirty-one thousand, two

You can also use place-value models to show place value.

Standard Form → 2,325

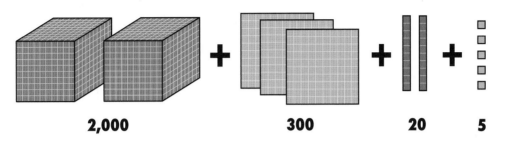

| 2,000 | 300 | 20 | 5 |

TRY THESE

Write each number in standard form.

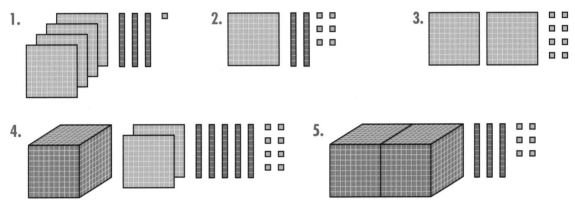

1.

2.

3.

4.

5.

Write in words and in standard form the values named by the following digits in the number 68,125,937.

6. 8   **7.** 2   **8.** 5   **9.** 7   **10.** 9

Copy each number and place commas to separate the periods. Read each number.

11. 8762   **12.** 90147   **13.** 807143   **14.** 7865319

Exercises

Write each number in standard form.

1. seven thousand, four hundred twenty-nine

2. two million, eight hundred sixty-five thousand, four hundred eleven

3. one hundred ninety-eight thousand, six hundred

4. one million, three hundred eighty thousand, fifty-five

5. 500,000 + 20,000 + 800 + 10 + 1

6. 3,000 + 90 + 2

7. 1,000,000 + 800,000 + 70,000 + 5,000 + 100 + 20 + 1

8. 4,000,000 + 800,000 + 2,000 + 700 + 30 + 7

PROBLEM SOLVING

9. What is the value of each 5 in the number 52,525? Write each value in standard form.

10. The Washington Monument in Washington, DC, has eight hundred ninety-eight steps. Write the number of steps in standard form.

11. The Sears Tower in Chicago, Illinois, is 1,450 feet tall. Write this number in expanded form.

12. The Eiffel Tower in Paris, France, is the most visited monument in the world. Approximately six million, four hundred twenty-eight thousand, four hundred people visit the tower each year. Write the number of people in standard form.

TEST PREP

13. Write the year you were born in standard form, expanded form, and words.

14. I am a four-digit number. My tens digit is four less than my hundreds digit. My thousands digit is 8. My ones digit is the smallest of my four digits. What number am I?

    a. 4,738   **b.** 8,621   **c.** 8,640   **d.** 8,959

Problem Solving

How We Spent Our Summer Vacation

Last summer, Gillian and Vicki had summer jobs. Gillian earned $380.35 the first month and $408.29 the second month. Vicki earned $399.17 the first month and $401.10 the second month.

- How much did each girl make for the two-month period?

- Who made the most money? How much more?

Extension

The girls want to buy a computer that costs $1,250. If they combine their money, will they have enough money to purchase the computer?

1.2 Numbers Through Hundred Millions

Objective: to extend place value through the hundred millions

Some countries have populations greater than a million. There are countries that have populations in the ten millions or hundred millions. In his research, Josh found a country with a population of approximately 136,212,921 people.

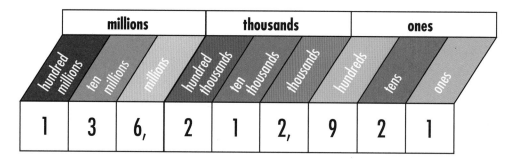

Standard Form → 136,212,921

Expanded Form → 100,000,000 + 30,000,000 + 6,000,000 + 200,000 + 10,000 + 2,000 + 900 + 20 + 1

Words ⟶ one hundred thirty-six million, two hundred twelve thousand, nine hundred twenty-one

TRY THESE

Write in words and in standard form the values named by the following digits in the number 136,212,921.

1. 6   **2.** 9   **3.** 3

Copy each number and place commas to designate the periods. Read each number.

4. 12453456   **5.** 19876330   **6.** 14440975

Write each number in standard form.

7. four hundred

8. two thousand, eight hundred seventy-eight

9. sixteen million, eight hundred thousand, four hundred twelve

10. one hundred million, nine hundred eighty-seven thousand, six hundred fifty-four

Exercises

Match the words to the correct standard form.

1. eight thousand, two hundred two

2. eight thousand, twenty-two

3. eighteen thousand, two hundred twenty-two

4. eight hundred twenty-two

5. eight hundred million, two hundred two thousand, twenty-two

6. eight thousand, two

7. eight million, two hundred thousand, twenty-two

a. 18,222

b. 8,202

c. 800,202,022

d. 8,200,022

e. 8,002

f. 8,022

g. 822

PROBLEM SOLVING

8. The population of Italy is approximately 58,103,457 people. Write the number in expanded form.

9. In 2006, the population of the United States was approximately 300,000,000 people. Write the population in words.

★10. Write all of the three-digit numbers possible using the digits 4, 6, and 8. Use each digit only once in each number.

★11. Write the current year in standard form, expanded form, and words.

MIXED REVIEW

Add or subtract mentally.

12. $1 + 3$

13. $5 + 2$

14. $2 + 7$

15. $3 + 4$

16. $6 + 0$

17. $10 - 3$

18. $5 - 3$

19. $8 - 6$

20. $6 - 1$

21. $9 - 5$

1.3 Different Names for a Number

Objective: to use place value to find equivalent names for numbers

In the United States, the president's house is called the Executive Mansion. It was painted white in 1814 to hide the marks from a fire during the War of 1812. That is why it also is known as the White House.

Numbers can have many names. When do you sometimes need to use another name for a number?

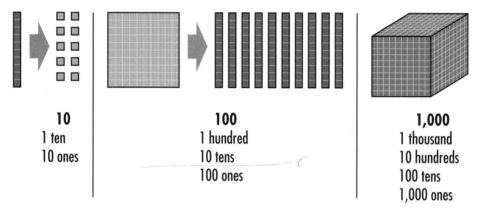

10
1 ten
10 ones

100
1 hundred
10 tens
100 ones

1,000
1 thousand
10 hundreds
100 tens
1,000 ones

More Examples

A. 70
   7 tens
   70 ones

B. 650
   65 tens
   650 ones

C. 2,600
   26 hundreds
   260 tens
   2,600 ones

D. 829,000
   829 thousands
   8,290 hundreds
   829,000 ones

TRY THESE

Choose another name for each number.

1. 43 tens   **a.** 43   **b.** 430   **c.** 4,300

2. 36 ones   **a.** 36   **b.** 360   **c.** 3,600

3. 9 tens   **a.** 9   **b.** 90   **c.** 900

4. 5 hundreds   **a.** 5   **b.** 50   **c.** 500

5. 62 thousands   **a.** 620   **b.** 6,200   **c.** 62,000

6. 88 millions   **a.** 88,000   **b.** 8,800,000   **c.** 88,000,000

Exercises

Write each number in standard form.

1. 8 hundreds   **2.** 5 tens   **3.** 7 thousands   **4.** 9 ones

5. 21 tens   **6.** 53 hundreds   **7.** 51 thousands   **8.** 67 hundreds

9. 230 thousands   **10.** 125 hundreds   **11.** 1,000 tens   **12.** 613 hundreds

13. 9 hundred millions   **14.** 56 millions

Copy and complete.

15. 20 = ■ ones   **16.** 400 = ■ tens   **17.** 3,000 = ■ hundreds

18. 200 = ■ hundreds   **19.** 7,000 = ■ thousands   **20.** 2,700 = ■ hundreds

21. 4,560 = ■ tens   **22.** 3,100 = ■ tens   **23.** 9,090 = ■ tens

24. 999,000 = 999 thousands or ■ hundreds

★**25.** 333,000,000 = 333 millions or ■ thousands

PROBLEM SOLVING

26. Mrs. Alvarez works at a bank. She has a stack of ten-dollar bills that equals $500. How many ten-dollar bills are in the stack?

27. Name another way to say 100,000.

    a. 1,000 hundreds

    b. 1,000 thousands

★**28.** The John Hancock Center in Chicago has 100 stories. Each story is about 10 feet high. Use place-value ideas to find the approximate height of the John Hancock Center. Explain.

★**29.** Find the number that does not belong.

    a. 1 million

    b. 10,000 hundreds

    c. 1,000,000

    d. 100 thousands

    e. 1,000 thousands

1.4 Comparing and Ordering

Objective: to compare and order whole numbers up to the millions place value

New York City, NY, and Los Angeles, CA, are two of the largest cities in the United States. New York has a population of approximately 15,626,754 people. Los Angeles has a population of approximately 15,329,817 people. Which city is larger?

To compare numbers, compare the digits in the same place-value position.

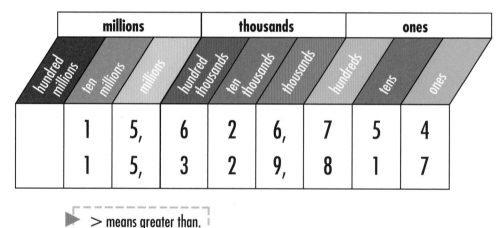

> > means greater than.
> < means less than.

Start with the greatest place-value position. Why?

- Compare the ten millions. The digits are the same.
- Compare the millions. The digits are the same.
- Compare the hundred thousands. 6 is greater than 3.
- 6 hundred thousands are greater than 3 hundred thousands.
- 15,626,754 is greater than 15,329,817.
- 15,626,754 > 15,329,817
- New York's population > Los Angeles' population

More Examples

A. 78 *is less than* 98.

    78 < 98

B. 177,405 *is equal to* 177,405.

    177,405 = 177,405

C. These numbers are in order from least to greatest.

    65   605   650   654

    65 < 605, 605 < 650, and 650 < 654

TRY THESE

Write the lesser number.

1. 94   99

2. 874   674

3. 5,823,646   4,285,646

4. 4,702   4,705

Write the greater number.

5. 346   354

6. 7,009,009   7,090,009

Exercises

Compare using <, >, or =.

1. 233 ● 238

2. 650 ● 560

3. 1,432 ● 1,432

4. 3,636 ● 3,686

5. 42,521 ● 42,521

6. 7,800,600 ● 7,800,200

7. 81,126,322 ● 81,045,567

8. 980,932,147 ● 908,932,147

Order the numbers from least to greatest.

9. 760   780   710

10. 520   537   525

11. 3,741   3,471   3,771

12. 62,051   60,051   62,501

13. 123,500 123,050 123,005

14. 491,000 490,001 490,000

15. 3,125,793 3,215,739 3,521,397

16. 82,127,644 8,212,764 282,712,464

PROBLEM SOLVING

17. Is three thousand, eighty-six greater than or less than three thousand, eighty-eight?

18. Put the following numbers in order from greatest to least.

5,246,321 5,246,362
5,246,307 5,246,400

MIXED REVIEW

Add or subtract mentally.

19. 7 + 5

20. 8 + 8

21. 6 + 6

22. 6 + 8

23. 5 + 9

24. 7 + 4

25. 4 + 9

26. 3 + 8

27. 7 + 7

28. 8 + 4

29. 12 − 7

30. 15 − 8

31. 18 − 9

32. 13 − 7

33. 12 − 4

34. 13 − 9

35. 11 − 7

36. 13 − 4

37. 11 − 5

38. 11 − 8

MID-CHAPTER REVIEW

Write each number in standard form.

1. five hundred forty

2. six thousand, seven hundred

3. 4,000,000 + 900,000 + 9,000 + 200 + 20 + 7

4. 88 million

5. 53 tens

6. 19 hundreds

7. 312 millions

8. 40 hundred thousands

Write the value of 5 in each number.

9. 159

10. 5,904

11. 64,759,981

12. 5,000,623

Write each number in standard form.

1.

2.

3.

4. eight hundred eleven

5. nine hundred ninety

6. two thousand, two hundred five

7. twelve thousand, twenty-four

8. three hundred sixty-six thousand

9. fifty-three thousand, thirty-one

10. four hundred million

11. seventeen million, five thousand

Write the value of 2 in each number.

12. 283   **13.** 4,020   **14.** 2,679   **15.** 218,510,341

Write each number in expanded form.

16. 406   **17.** 12,050   **18.** 706,000   **19.** 800,450,120

Write each number in words.

20. 650   **21.** 2,028   **22.** 14,500   **23.** 6,400,000

Copy and complete.

24. 30 = ■ ones   **25.** 870 = ■ tens   **26.** 6,200 = ■ hundreds

27. 444 = ■ ones   **28.** 345,986,500 = ■ hundreds   **29.** 7,300 = ■ tens

Solve.

30. Find the number. The hundreds digit is 6. The ones and tens digits are the same. The ones digit plus the tens digit equals the hundreds digit.

31. The Grand Canyon is two hundred seventeen miles long. Write this number in standard form. Use place-value models to show the number.

Objective: to solve problems using the four-step plan

There are about 82,312,554 people living in Germany. There are about 61,243,546 people living in France. How many more people live in Germany than in France?

You can use a **four-step plan** to help you solve problems.

1. READ

Read the problem carefully.

What do you know? the number of people living in Germany and France; there are more people living in Germany

What do you need to find out? the difference between the number of people living in each country

2. PLAN

Decide how to solve the problem.

Choose a strategy to help you. Since you want to know the difference between the populations of the two countries, subtract.

Write an equation like this to help you:

82,312,554 – 61,243,546 = ▪

3. SOLVE

Compute or work through the plan.

82,312,554 – 61,243,546 = 21,069,008 people

Be sure to label the solution.

4. CHECK

Reread the problem.

Check your solution.

Add the difference (the solution) to the population of France. This will give you the population of Germany.

61,243,546 + 21,069,008 = 82,312,554 people ✓

TRY THESE

1. The population of St. Petersburg, Florida, is about 249,079 people. The population of St. Petersburg, Russia, is about 5,132,072 people. What is the total population of both cities?

2. The Empire State building is 1,250 feet tall. The Leaning Tower of Pisa is about 180 feet tall. Which tower is shorter? How much shorter is it?

1. Isabel has 14 stickers. Tracy has 5 times as many stickers as Isabel. How many stickers does Tracy have?

2. Luke has 1,281 baseball cards. Simon has 1,535 baseball cards. If Luke and Simon combine their baseball cards, about how many cards will they have altogether: 2,500 or 3,000? Estimate.

★ 3. There are 369 families who each donated $6.00 to the Calvert fundraiser. How much money was donated?

MIND BUILDER

Logical Reasoning: Vacation Time

Use logic to determine where each of these four friends is going on vacation. Consult the clues and a map and make a table like this one to help you. Each friend is going to only one place, and no two of them are going to the same place.

	England	Venezuela	China	Mexico
Fran				
Dan		No	No	
Stan				
Nan				

1. Dan is not going to South America or Asia.

2. Either Dan or Nan is going to Europe.

3. Fran is going to either a North or South American country.

4. Stan knows Spanish, but it isn't going to help him where he is going.

5. Nan is going to see Big Ben and Buckingham Palace.

6. One of Fran's friends is going to see the Great Wall.

1.6 Constructing a Constructed Response

Objective: to communicate mathematical thinking in writing

During geography class, Mr. Edwards placed cards on the board with the approximate 2007 urban populations of the world's ten largest cities. He challenged the students to study the board to find the largest city in the world. Paul said that it was Mexico City. Maria claimed that it was New York City. Eliza disagreed with them both and thought Tokyo was the largest city. How can you determine who was right? Explain your reasoning.

New York City, USA
19,040,000 people

Buenos Aires, Argentina
12,795,000 people

Tokyo, Japan
35,676,000 people

Mexico City, Mexico
19,028,000 people

Mumbai, India
18,978,000 people

Delhi, India
15,926,000 people

São Paulo, Brazil
18,845,000 people

Shanghai, China
14,987,000 people

Calcutta, India
14,787,000 people

Dhaka, Bangladesh
13,485,000 people

Sometimes with math problems, finding the correct answer is only half of what the question is asking. The question also can ask you to explain *how* you found that answer or how you know your answer is right. This type of question is called a **constructed response**. When you answer a constructed response question, you need to write your thoughts on paper to show your thinking.

You can use the four-step plan to help you solve problems.

1. READ
You know that Paul, Maria, and Eliza all have different opinions about what is the largest city. You need to find out who is right.

2. PLAN
Think about place value. Compare the ten millions of each number to find the largest number.

Tokyo	35,676,000
New York City	19,040,000
Mexico City	19,028,000
Mumbai	18,978,000
São Paulo	18,845,000
Delhi	15,926,000
Shanghai	14,987,000
Calcutta	14,795,000
Dhaka	13,485,000
Buenos Aires	12,795,000

3. SOLVE
Make a list to put the cities in order by looking at the ten millions and one millions place values. You may need to use the hundred thousands and ten thousands. Then find the largest city.

Since Tokyo has the most ten millions, it is the city that has the greatest population. Eliza was right.

4. CHECK
Does your answer make sense? yes

You can check it by making sure you read your numbers correctly and placed all of the cities in the proper order.

In the answer to a constructed response question, you write how you solved the problem, including all of your reasoning. You should also include how you checked your work.

A good constructed response to the preceding problem might look like this:

> I used place value to compare the numbers. I then ordered the cities in a list from largest to smallest. The list made it easy to see that Tokyo is the largest city because Tokyo's population had the most ten millions. Neither Paul nor Maria were correct. Maria came closer with New York City, which was the second highest. Paul's choice of Mexico City is smaller than New York City and Tokyo. Eliza was correct to disagree with Paul and Maria and choose Tokyo.

When writing constructed responses, be certain to solve the problem, write a description of your solution, and defend your position against any others.

TRY THESE

1. Arrange the digits 5, 3, 1, 7, 2, 8, and 9 into the smallest seven-digit number you can. Use each digit only once. Write the number, and explain how you know it is the smallest number you can make?

2. A hippo at the zoo can eat 90 pounds of grass per day. At this rate, how much will this hippo eat in a week? Explain how you know that your answer is correct.

Solve

Use the chart to solve problems 1–5.

1. David was studying a book about world languages and made this chart showing the different languages that people speak on Earth. About how many more people worldwide speak Russian than Arabic? Explain how you found your answer?

2. David imagined that within the next century the number of speakers for each language would grow by 1 million people. If that was true, how many people would speak Spanish in the next century? Explain your reasoning.

World Languages

Language	Number of People
Arabic	265,284,901
Bengali	245,208,058
Chinese Mandarin	1,246,345,092
English	512,827,267
French	77,208,105
German	100,082,393
Hindi	501,207,974
Japanese	125,488,387
Korean	78,297,033
Malay-Indonesian	140,497,007
Portuguese	196,697,266
Russian	285,284,998
Spanish	399,029,398

3. David was curious about the number of people who speak French, English, Japanese, and Korean. He put them in order from least to greatest to see which of the four had the greatest population of speakers. Create his list.

4. David wants to compare the language with the smallest number of speakers to the language that has the largest number. What are these languages? What is the difference in the number of speakers of each language?

5. English and German are known as Germanic languages. According to David's findings, what is the total number of speakers of Germanic languages?

MIXED REVIEW

Complete.

6. Write 782,091 in expanded form.

7. Complete the sentence. Each group of three digits, starting from the ones place, is called a _____.

8. *True* or *False*: In the number 9,683,275, the 3 has a value of 1,000.

9. Place the commas in the following number: 34211092.

10. Write the standard form of the number shown by the model.

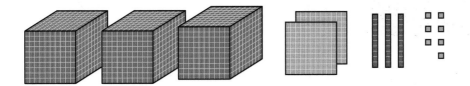

MIND BUILDER

Patterns

Several periods greater than millions are listed to the right.

Solve.

Period	Standard Form	Number of Zeros
Billion	1,000,000,000	9
Trillion	1,000,000,000,000	12
Quadrillion	1,000,000,000,000,000	15

1. Write sixty-three billion in standard form.

2. Which is greater, thirteen trillion or thirteen quadrillion?

3. Write four quadrillion, three hundred twenty-six trillion in standard form.

1.7 Rounding

Objective: to round whole numbers through the millions

Adam's Peak is a mountain located in southwestern Sri Lanka. It has an elevation of about 7,360 ft.

You can round numbers without a number line. Round 7,360 to the nearest thousand.

1. Circle the digit being rounded.

2. Look at the digit to the right of the circled digit.

3. If the digit to the right is 5 or greater, round up.

4. If the digit to the right is 0, 1, 2, 3, or 4, round down.

⑦,360
⑦,360
↓
7,000

▶ The circled digit remains the same if the digit to the right is 0, 1, 2, 3, or 4.

To the nearest thousand, 7,360 rounds to 7,000.

More Examples

A. Round 26,512 to the nearest thousand.

2⑥,512   Circle the 6. Look at the digit to the right.
↓
②7,000   The digit is 5. Round up.

B. Round 8,067,457 to the nearest hundred thousand.

8,⓪67,457
↓   6 > 5 Round up.
8,100,000

C. Round 37,431,250 to the nearest million.

3⑦,431,250
↓   4 < 5 Round down.
37,000,000

TRY THESE

Choose the correct answer for rounding to the nearest ten.

1. 46   **a.** 40   **b.** 50   **2.** 63   **a.** 60   **b.** 70

3. 374   **a.** 370   **b.** 380   **4.** 435   **a.** 430   **b.** 440

Exercises

Round to the nearest hundred.

1. 610   **2.** 640   **3.** 382   **4.** 570   **5.** 851   **6.** 639

7. 1,320   **8.** 4,866   **9.** 2,393   **10.** 7,458   **11.** 1,708   **12.** 2,959

Round to the nearest thousand.

13. 8,100   **14.** 6,089   **15.** 2,875   **16.** 4,580   **17.** 3,709

Round to the nearest million.

18. 7,999,672   **19.** 3,388,431   **20.** 20,672,590   **21.** 75,400,760

22. 49,551,557   **23.** 89,869,485   **24.** 1,303,321   **25.** 279,562,777

26. 91,234,567   **27.** 1,876,000   **28.** 34,567,324

PROBLEM SOLVING

29. The Superdome in New Orleans has a maximum football seating capacity of 72,003. To the nearest thousand, about how many people can be seated at a football game?

30. The Rose Bowl in California has 92,542 seats. To the nearest hundred thousand, about how many persons can be seated in the Rose Bowl?

31. Michigan Stadium has 107,501 seats. To the nearest thousand, about how many persons can be seated in Michigan Stadium?

TEST PREP

32. 4,317 rounded to the nearest hundred is _____.

   a. 5,000   **b.** 4,320   **c.** 4,400   **d.** 4,300

33. 72,430 rounded to the nearest thousand is _____.

   a. 72,000   **b.** 73,000   **c.** 80,000   **d.** 70,000

LANGUAGE and CONCEPTS

Write the letter of the word or number that best completes each sentence.

1. Five hundred six written in standard form is _____.

2. The standard form for 5,000 + 60 is _____.

3. Another name for 56 hundreds is _____.

4. 9,050 is _____ than 9,051.

5. Rounded to the nearest million, 58,987,564 is _____.

6. Rounded to the nearest ten, 564 is _____.

a. 506
b. 560
c. 59,000,000
d. 5,060
e. 5,600
f. greater
g. less

SKILLS and PROBLEM SOLVING

Write each number in standard form. (Sections 1.1–1.2)

7. seven hundred

8. nine hundred forty

9. four hundred fifty-nine

10. six hundred million, five hundred thirty-two thousand, fourteen

11. thirteen thousand

12. fifty thousand, four hundred

13. six hundred thousand

14. 53 million, 289 thousand

Write each number in words. (Sections 1.1–1.2)

15. 450,921,374   16. 920   17. 18,700   18. 5,000,000

Write each number in expanded form. (Sections 1.1–1.2)

19. 222   20. 5,600   21. 19,400   22. 283,916,413

Write the value of each 4. (Sections 1.1–1.2)

23. 742   24. 8,403   25. 451,666   26. 4,756,802

Choose another name for each number. (Section 1.3)

27. 53 tens   **a.** 53   **b.** 530   **c.** 5,300

28. 6 hundreds   **a.** 6   **b.** 60   **c.** 600

29. 75 hundreds   **a.** 7,500   **b.** 75,000   **c.** 750

30. 14 thousands   **a.** 1,400   **b.** 14,000   **c.** 140,000

Compare using <, >, or =. (Section 1.4)

31. 281,327,405 ● 281,327,405

32. 752,483 ● 725,483

33. 17,903,618 ● 18,903,826

Round to the nearest ten million. (Section 1.7)

34. 53,235,612   **35.** 67,438,099   **36.** 366,264,531

37. 275,379,611   **38.** 892,432,095

Solve. (Sections 1.5–1.6)

39. Find the number. The thousands digit is 6. The tens digit is 8. The tens digit is one less than the hundreds digit. The ones digit is 3 more than the thousands digit.

40. Find the number. The tens digit is 3. The ones digit is 2 less than the tens digit. The thousands digit is six more than the ones digit. The hundreds digit is 3 less than the tens digit.

41. Explain how you found the answer to problem 40.

CHAPTER 1 TEST

Write each number in standard form.

1. seven hundred five

2. five thousand, seventy-one

3. forty-four thousand, forty-four

4. nineteen million

Write the value of each 8.

5. 5,824   **6.** 8,706   **7.** 875,321   **8.** 18,000,947

Write each number in words.

9. 700   **10.** 3,004   **11.** 150,000   **12.** 12,600,000

Copy and complete.

13. 500 = ■ tens   **14.** 74,000,000 = ■ millions   **15.** 5,640 = ■ tens

Compare using <, >, or =.

16. 145 ● 148   **17.** 1,375 ● 1,378   **18.** 13,200 ● 13,199

Round to the nearest hundred.

19. 485   **20.** 4,428   **21.** 653   **22.** 21,136

Solve.

23. Debbie went on a trip to the moon. She traveled 238,854 miles from Earth to the moon. Donna went on a trip to the sun. She traveled 93,213,481 miles from Earth to the sun.

  a. Use <, >, or = to compare the distances.

  b. Round each number to the nearest hundred thousand.

  c. How much farther did Donna travel than Debbie?

24. By the time Debbie and Donna returned from their trips, it was the year 2215. Write the year in expanded form and in words.

Egyptian Numbers

Different cultures use different symbols for numbers. The Romans used letters to represent their numbers. The Egyptians used the symbols on the right to represent numbers.

What number is named by the Egyptian numeral?

1.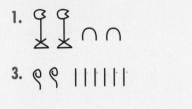

2. ﹒﹒﹒

3. ﹒﹒﹒ ||||||

4. ﹒﹒﹒

5.

Change to an Egyptian numeral. Use the chart.

6. 314

7. 8,234

8. 12,345

9. 20,345

★ 10. 1,111,111

Egyptian Numbers

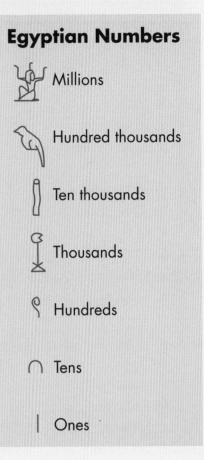

Millions

Hundred thousands

Ten thousands

Thousands

Hundreds

Tens

Ones

Solve.

11. What is your age in Egyptian numerals?

12. Pyramids are tombs built for the kings. In Egypt, the largest pyramid is called the Great Pyramid. The Great Pyramid was built for the great king, King Khufu. This pyramid took almost 30 years to build. Write the number of years in Egyptian numerals.

13. King Tutankhamun's tomb was found in 1922. Write the year in Egyptian numerals.

14. The first pyramid was the Step Pyramid at Saqqara, built for King Zoser in 2750 B.C. Write the year in Egyptian numerals.

CUMULATIVE TEST

1. What is the standard form of twenty thousand, forty-seven?
   a. 2,047
   b. 20,047
   c. 20,407
   d. none of the above

2. Which number is equal to 500,000,000 + 200,000 + 70,000 + 3,000 + 900?
   a. 500,237,900
   b. 500,273,900
   c. 500,293,700
   d. none of the above

3. Which digit is in the thousands place in 68,192?
   a. 6
   b. 8
   c. 9
   d. none of the above

4. What is another name for 53 hundreds?
   a. 53
   b. 530
   c. 5,300
   d. none of the above

5. Which number is greater than 28,517,623?
   a. 9,999,999
   b. 28,463,543
   c. 28,520,007
   d. none of the above

6. What is 42,509 rounded to the nearest thousand?
   a. 42,000
   b. 42,500
   c. 42,510
   d. none of the above

7. Round to the nearest hundred to find the approximate depth of the Salton Sea.
   a. 1,000 cm
   b. 1,400 cm
   c. 1,500 cm
   d. 2,000 cm

Lake	Depth in Centimeters
Wawasee	2,073
Salton Sea	1,463
Oneida	1,524
Moosehead	7,498

8. The chart shows the approximate population of each continent in 2006. Which continent has more people than North America but less than Africa?
   a. Australia
   b. Asia
   c. Europe
   d. South America

Continents	Populations
Africa	877,500,000
Antarctica	0
Australia	32,000,000
Asia	3,879,000,000
Europe	727,000,000
North America	501,500,000
South America	379,500,000

Adding and Subtracting Whole Numbers

Sarah Mesrobian
Virginia

2.1 Properties of Addition

Objective: to identify and use the properties of addition

Cindy and Eric went bowling. Cindy knocked down 6 pins with her first ball. She knocked down 3 pins with the second ball. Eric first knocked down 3 pins, then 6 pins. They each have a score of 9 points.

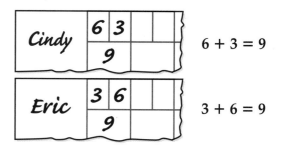

A. Commutative Property of Addition

Changing the order of the **addends** does not change the **sum**.

7 + 2 = 9
2 + 7 = 9

B. Identity Property of Addition

When 0 is added to any number, the sum is that number.

0 + 5 = 5
2 + 0 = 2

C. Associative Property of Addition

Grouping the addends in a different way does not change the sum.

(3 + 4) + 2 = 3 + (4 + 2)
7 + 2 = 3 + 6
9 = 9

▶ Parentheses tell which numbers to add first.

More Examples

A. 6 + 8 = 14

8 + 6 = 14

B. 50 + 0 = 50

C. (15 + 10) + 5 = 15 + (10 + 5)

25 + 5 = 15 + 15

30 = 30

TRY THESE

Name the property shown.

1. 100 + 0 = 100

2. 24 + 36 = 60
36 + 24 = 60

3. 5,678 + 0 = 5,678

4. (22 + 6) + 4 = 32
22 + (6 + 4) = 32

Exercises

Add. Use the properties of addition to help you.

1. $14 + 6 + 2$ **2.** $12 + 8$ **3.** $30 + 12 + 0$ **4.** $71 + 0$

5. $8 + 32 + 40$ **6.** $11 + 9$ **7.** $0 + 152$ **8.** $60 + 14 + 2$

9. $10 + 9$ **10.** $0 + 10 + 9$ **11.** $3 + 50 + 2$ **12.** $60 + 2 + 14$

13. $103 + 21 + 6$ **14.** $90 + 25 + 20$ **15.** $340 + 6 + 4$

Copy and complete.

16. $8 + 9 = 9 + ■$ **17.** $■ + 0 = 6$ **18.** $(7 + 3) + 4 = 7 + (3 + ■)$

19. $(10 + 20) + 30 = ■ + (20 + 30)$ **20.** $300 + 400 = ■ + 300$

21. $53 + 89 = 89 + ■$ **★22.** $12 + (15 + 19) = 12 + (19 + ■)$

23. Use the following numbers to write two addition equations.

   7   9   16

PROBLEM SOLVING

24. Name the property shown in following problem.
   $(4 + 6) + 8 = 4 + (6 + 8)$

25. Does the Commutative Property apply when one of the addends is 0?

26. Connie's Pet Store has 67 pets. It has 31 dogs, 14 rabbits, and the rest are cats. How many cats does the pet store have?

CONSTRUCTED RESPONSE

27. Cindy's and Eric's bowling scores are shown. Eric's total score is 198. What is Cindy's total score? Explain how you know.

	Game 1	Game 2
Cindy	105	93
Eric	93	105

2.2 Expressions and Equations

Objective: to distinguish between a mathematical expression and an equation

Kay and her brother biked 8 miles before lunch and 9 miles after lunch. $8 + 9$ represents the miles they biked for the total trip. $8 + 9$ is a mathematical expression. A **mathematical expression** is a translation of a word phrase or fragment.

Phrase	Expression
the sum of 12 and 2 two more than 12	$12 + 2$
the difference of 9 and 7 9 less 7	$9 - 7$
2 times 4 the product of 2 and 4	2×4
16 divided by 8 the quotient of 16 and 8	$16 \div 8$

You can translate word sentences into mathematical sentences. A mathematical sentence with an equal sign is an **equation**.

Sentence	Equation
The sum of 6 and 12 is equal to 18.	$6 + 12 = 18$
7 less 5 is equal to 2.	$7 - 5 = 2$
The product of 6 and 8 is 48.	$6 \times 8 = 48$
14 divided by 7 is 2.	$14 \div 7 = 2$

Operation symbols ($+$, $-$, $\times$, $\div$) are used to write both mathematical expressions and equations. The four basic operations are addition, subtraction, multiplication, and division.

TRY THESE

Translate each of the following. Tell if it is an expression or an equation.

1. 13 minus 4

2. 7 plus 9 is equal to 16.

3. 9 multiplied by 2 is equal to 18.

4. the quotient of 24 and 8

Exercises

Translate each of the following. Tell if it is an expression or an equation.

1. 3 less than 10 is 7.

2. the sum of 7 and 11

3. fifteen divided by 3

4. The product of 3 and 9 is 27.

5. 13 times 8

6. The quotient of 25 and 5 is 5

7. The total of 21 and 25 is 46.

8. 20 decreased by 12

9. Forty divided by ten is four.

10. the sum of 9,000 and 21

11. 11 multiplied by 3 is 33.

12. 10 less 8 is 2.

13. The product of 12 and 4 is 48.

14. the difference of 84 and 14

15. The sum of 81 and 13 is 94.

16. Twelve divided by two

PROBLEM SOLVING

17. Sam has 81 plastic blocks. Lee gives him a box with 220 plastic blocks in it for his birthday. How many plastic blocks does Sam have now?

18. Sean sees 14 geese swimming in the water and 38 geese flying in the sky. How many geese does Sean see altogether?

19. Kelly collects stuffed animals. She has a total of 117 stuffed animals. She has 34 bears, 17 dogs, 13 cats, and all the rest are horses. How many stuffed horses does Kelly have?

TEST PREP

20. 987,000 + 3,000

a. 100,000,000

b. 99,000,000

c. 990,000

d. none of the above

2.3 Computation with Parentheses

Objective: to simplify math expressions and solve equations that contain parentheses

Doug cooked 6 hot dogs and 7 hamburgers.

If 8 sandwiches were eaten, how many were left?

You add to find the total number of sandwiches. Then subtract the number of sandwiches eaten.

$(6 + 7) - 8$

$13 \quad - 8 = 5$

▶ Work the operation inside the parentheses first.

There were 5 sandwiches left.

More Examples

A. $12 - (3 + 4) = \blacksquare$

$12 - \quad 7 \quad = 5$

B. $(15 - 8) + 4 = \blacksquare$

$7 \quad + 4 = 11$

C. $(40 + 50) - 20 = \blacksquare$

$90 \quad - 20 = 70$

TRY THESE

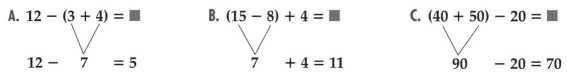

Match.

1. $(12 - 6) + 3$ **a.** $7 + 7$ **5.** $(8 + 1) - 7$ **e.** $15 - 4$

2. $7 + (9 - 2)$ **b.** $80 - 50$ **6.** $17 - (4 + 5)$ **f.** $130 - 50$

3. $(2 + 9) - 4$ **c.** $11 - 4$ **7.** $15 - (8 - 4)$ **g.** $9 - 7$

4. $80 - (30 + 20)$ **d.** $6 + 3$ **8.** $(40 + 90) - 50$ **h.** $17 - 9$

Exercises

Simplify.

1. $(2 + 5) - 4$
2. $16 - (4 + 4)$
3. $(15 - 9) + 8$
4. $(13 + 3) - 7$
5. $17 - (6 + 2)$
6. $(18 - 9) + 7$
7. $(50 + 10) - 40$
8. $(60 - 20) + 40$
9. $20 + (50 - 40)$
10. $120 - (40 + 20)$
11. $(90 + 60) - 80$
★ 12. $(200 + 300) - 400$
★ 13. $800 - (300 + 100 + 400)$

Complete.

★ 14. In $15 - (3 + 4)$, do you *add* or *subtract* first?

★ 15. Do the following two problems have the same answer? Explain.
$(9 - 4) + 3 = \blacksquare$
$9 - (4 + 3) = \blacksquare$

16. Jacob earned $9 mowing lawns and $7 painting a fence. He spent $8 for a tire for his bicycle. How much money does he have left to put in the bank?

17. To reach the cabin, Nathan traveled 17 miles. He canoed 6 miles and hiked 2 miles. He drove the rest of the way. How far did he drive?

★ 18. Place parentheses in the following problem so that the answer is 10.
$70 - 20 + 40 = 10$

MIND BUILDER

Palindromes

A number that uses the same digits forward and backward is called a **palindrome**.

To find the palindrome for a number, do the following:

- Write a number.
- Reverse the digits.
- Add. Is the sum a palindrome?
- If not, reverse the digits and add again until the sum is a palindrome.

$$
\begin{array}{r}
57 \\
+ \ 75 \\
\hline
132 \\
+ \ 231 \\
\hline
363
\end{array}
$$

363 is a palindrome.

Find the palindrome for each number.

1. 32
2. 243
3. 3,712
4. 911
5. 8,367
6. 256

2.4 Estimating Sums and Differences

Objective: to estimate sums and differences

There were 484 people at the 4-H cattle sale in the morning and 337 people in the afternoon. About how many more people were at the sale in the morning?

When you only need to know *about* how many, **estimate**.

To estimate 484 − 337, round each number to the highest place value they both share. In this case, round to the nearest hundred.

$$
\begin{array}{r}
484 \rightarrow 500 \\
- 337 \rightarrow - 300 \\
\hline
200
\end{array}
$$

484 rounds to 500.
337 rounds to 300.
Estimate.

There were *about* 200 more people at the sale in the morning.

More Examples

Estimate sums and differences as follows.

- Round each number to the greatest place-value position that both numbers share.
- Then add or subtract.

A.
$$
\begin{array}{r}
38 \rightarrow 40 \\
+ 54 \rightarrow + 50 \\
\hline
90
\end{array}
$$
Round to the nearest ten.

B.
$$
\begin{array}{r}
2,749 \rightarrow 3,000 \\
+ 3,826 \rightarrow + 4,000 \\
\hline
7,000
\end{array}
$$
Round to the nearest thousand.

C.
$$
\begin{array}{r}
74 \rightarrow 70 \\
- 33 \rightarrow - 30 \\
\hline
40
\end{array}
$$

D.
$$
\begin{array}{r}
675 \rightarrow 700 \\
+ 521 \rightarrow + 500 \\
\hline
1,200
\end{array}
$$

E.
$$
\begin{array}{r}
3,256 \rightarrow 3,000 \\
- 1,986 \rightarrow - 2,000 \\
\hline
1,000
\end{array}
$$

TRY THESE

Estimate by rounding.

1.
$$
\begin{array}{r}
54 \rightarrow 50 \\
+ 37 \rightarrow + 40
\end{array}
$$

2.
$$
\begin{array}{r}
918 \rightarrow 900 \\
- 485 \rightarrow - 500
\end{array}
$$

3.
$$
\begin{array}{r}
3,254 \rightarrow 3,000 \\
+ 4,785 \rightarrow + 5,000
\end{array}
$$

4.
$$
\begin{array}{r}
76 \\
- 41
\end{array}
$$

5.
$$
\begin{array}{r}
721 \\
+ 455
\end{array}
$$

6.
$$
\begin{array}{r}
367 \\
+ 208
\end{array}
$$

7.
$$
\begin{array}{r}
6,314 \\
+ 7,288
\end{array}
$$

8.
$$
\begin{array}{r}
8,562 \\
- 1,241
\end{array}
$$

1. 29
   + 44

2. 94
   − 39

3. 274
   + 432

4. 896
   − 148

5. 5,372
   + 7,114

6. 8,965
   − 5,897

7. 829
   − 675

8. 8,115
   + 6,683

9. 482
   − 241

10. 7,753
   − 3,276

11. 813 − 548

12. 173 + 124

13. 6,784 − 2,941

14. 49 + 35 + 86

15. 542 + 301 + 726

★**16.** 2,324 + 3,616 + 1,777

★ **17.** One addend rounds to 500. The other addend rounds to 400. What is the estimated sum?

★ **18.** *Estimate* how much less 68 is than 94.

PROBLEM SOLVING

19. *About* how many total cattle are to be sold?

Cattle to be Sold	
Dairy	423
Beef	758

20. *Estimate* the total amount of money in the banks.

$36 $52

21. There are 438 sheep and 178 pigs in the barn. About how many more sheep are in the barn than pigs?

22. Make up a problem that can be solved by using this estimate. 40 + 70 = 110

MIXED REVIEW

Write each number in standard form.

23. eight hundred nine

24. thirty-four thousand, fifty-two

25. three hundred fifty thousand

26. twelve million

What is the value of each 6?

27. 869

28. 7,256

29. 18,604

30. 360,408

2.4 Estimating Sums and Differences   35

2.5 Adding Greater Numbers

Objective: to add greater numbers

Ben Winters helps feed the farm animals at the fair. He mixes 842 pounds of oats with 476 pounds of bran. How many pounds of feed does he have?

Add 842 and 476 to find the amount of feed.

THINK An estimate is 800 + 500 = 1,300.

Step 1	Step 2	Step 3
Add the ones.	Add the tens.	Add the hundreds.
$$\begin{array}{r} 842 \\ +\ 476 \\ \hline 8 \end{array}$$	$$\begin{array}{r} ^{1} \\ 842 \\ +\ 476 \\ \hline 18 \end{array}$$	$$\begin{array}{r} ^{1} \\ 842 \\ +\ 476 \\ \hline 1,318 \end{array}$$
No regrouping is needed.	Regroup 11 tens as 1 hundred 1 ten.	Regroup 13 hundreds as 1 thousand 3 hundreds.

Ben has 1,318 pounds of feed.

Is the answer reasonable? Compare it to the estimate.

More Examples

A.
$$\begin{array}{r} ^{1} \\ \$67 \\ +\ \ 45 \\ \hline \$112 \end{array}$$

B.
$$\begin{array}{r} ^{1\ 1} \\ 476 \\ +\ 754 \\ \hline 1,230 \end{array}$$

C.
$$\begin{array}{r} ^{1\ 1} \\ 3,757 \\ +\ \ \ 67 \\ \hline 3,824 \end{array}$$

D.
$$\begin{array}{r} ^{1\ 1} \\ \$56.48 \\ +\ 24.81 \\ \hline \$81.29 \end{array}$$

E.
$$\begin{array}{r} ^{1\ 22} \\ 567 \\ 89 \\ +\ 3,895 \\ \hline 4,551 \end{array}$$

TRY THESE

Add. Check your answer by estimation.

1.
$$\begin{array}{r} 78 \\ +\ 46 \end{array} \rightarrow \begin{array}{r} 80 \\ +\ 50 \\ \hline 130 \end{array}$$

2.
$$\begin{array}{r} 96 \\ +\ 75 \end{array} \rightarrow \begin{array}{r} 100 \\ +\ 80 \\ \hline 180 \end{array}$$

3.
$$\begin{array}{r} 297 \\ +\ 86 \end{array} \rightarrow \begin{array}{r} 300 \\ +\ 90 \\ \hline 390 \end{array}$$

4.
$$\begin{array}{r} 58 \\ +\ 67 \end{array}$$

5.
$$\begin{array}{r} 85 \\ +\ 89 \end{array}$$

6.
$$\begin{array}{r} \$438 \\ +\ \ 95 \end{array}$$

7.
$$\begin{array}{r} 676 \\ +\ 86 \end{array}$$

8.
$$\begin{array}{r} \$4.59 \\ +\ 0.76 \end{array}$$

Add.

1. 567
    433
    + 625

2.   29
    859
    + 472

3. 789
     52
    + 6,388

4.   43
    251
    + 2,704

5.   326
    5,271
    + 151

6. 8,973
    +   53

7. $23.86
    + 7.06

8.   458
    + 7,925

9. 6,491
    + 267

10. $14.99
    + 3.25

11. 774 + 5,985

12. 3,063 + 972

13. $7,127 + 1,823

14. 697 + 524 + 202 + 674

15. 21 + 36 + 43 + 54

16. $67 + $45

17. $76 + $24

18. $113 + $457

Find the missing addend.

19. 79 + ■ = 122

20. ■ + 551 = 1,427

★ 21. $5.98 + ■ = $9.88

Simplify.

★ 22. (876 + 551) − 100

★ 23. 700 − (32 + 25 + 43)

PROBLEM SOLVING

Use the graph to solve.

24. Are there more elephants or cheetahs?

25. How many cheetahs and lions are at the zoos?

26. How many flamingos are at the zoos?

27. What is the total number of giraffes, elephants, and lions?

★ 28. If 35 lionesses had 2 cubs each, how many more lions would there be? What would be the total number of lions?

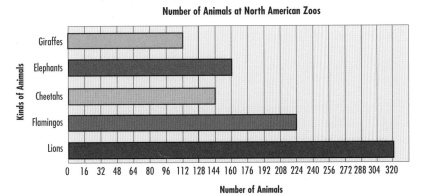

Number of Animals at North American Zoos

2.6 Problem-Solving Strategy: Choosing a Reasonable Answer

Objective: to determine a reasonable solution for a problem

Marcia's entire backyard is a gigantic flower garden. One part of it has 300 roses. She has other flowers in her garden, too. About how many flowers does she have in her garden?

Which answer makes sense?   **a.** 30   **b.** 300   **c.** 3,000   **d.** 30,000

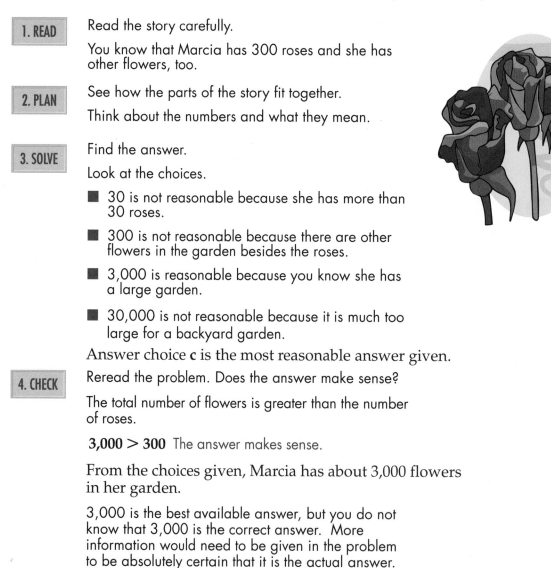

1. READ

Read the story carefully.

You know that Marcia has 300 roses and she has other flowers, too.

2. PLAN

See how the parts of the story fit together.

Think about the numbers and what they mean.

3. SOLVE

Find the answer.

Look at the choices.

■ 30 is not reasonable because she has more than 30 roses.

■ 300 is not reasonable because there are other flowers in the garden besides the roses.

■ 3,000 is reasonable because you know she has a large garden.

■ 30,000 is not reasonable because it is much too large for a backyard garden.

Answer choice **c** is the most reasonable answer given.

4. CHECK

Reread the problem. Does the answer make sense?

The total number of flowers is greater than the number of roses.

3,000 > 300 The answer makes sense.

From the choices given, Marcia has about 3,000 flowers in her garden.

3,000 is the best available answer, but you do not know that 3,000 is the correct answer. More information would need to be given in the problem to be absolutely certain that it is the actual answer.

TRY THESE

Choose the most reasonable answer. Tell why it makes sense.

1. About how many seconds does it take you to count to 100?

   a. 6   **b.** 60   **c.** 6,000

2. The math club had a car wash to raise money. How much did they charge to wash a car?

   a. 3¢   **b.** $3   **c.** $30

Choose the most reasonable answer.

1. Sharon runs 100 meters in 18 seconds. Gus takes less time. How many seconds does it take Gus?

   a. 16   **b.** 18   **c.** 20

2. Seth earned 7 points for diving. Lisa earned more. How many points did Lisa earn?

   a. 5   **b.** 7   **c.** 9

3. Martin has nineteen rose bushes in his garden. Five are red. How many bushes are not red?

   a. 4   **b.** 5   **c.** 14

4. Manuel said he has a younger sister and an older brother. How old is Manuel?

   a. 14 days   **b.** 14 weeks   **c.** 14 years

CONSTRUCTED RESPONSE

5. Look at **Solve** problem 4. How do you know that your answer makes the most sense? Explain your reasoning.

MID-CHAPTER REVIEW

Copy and complete.

1. $7 + 6 = \blacksquare + 7$

2. $4 + \blacksquare = 4$

3. $6 + (2 + 3) = (6 + 2) + \blacksquare$

Find the missing addend.

4. $5 + \blacksquare = 13$

5. $\blacksquare + 7 = 15$

6. $6 + \blacksquare = 14$

Estimate.

7. $\begin{array}{r} 43 \\ -\ 28 \end{array}$

8. $\begin{array}{r} 82 \\ +\ 36 \end{array}$

9. $\begin{array}{r} 354 \\ -\ 492 \end{array}$

Add.

10. $\begin{array}{r} 729 \\ +\ 533 \end{array}$

11. $\begin{array}{r} 3,841 \\ +\ 5,208 \end{array}$

Problem Solving

Weird Wacco's Quiz Show

Penny Packer appeared on Weird Wacco's Quiz Show. She had to pick four numbers that used three steps to get to a common difference. If she could do this, she would win a prize.

Weird Wacco asked Penny Packer to choose four numbers and write them in a square.

She then connected two of the numbers without going across the middle. She found the difference between the two numbers and wrote it at the center of the line.

Weird Wacco told Penny Packer to do the same with the other pairs.

Penny Packer then had to connect these numbers with lines and find their differences.

Penny Packer won!

For the grand prize, Penny has to pick four numbers that need four steps. Weird Wacco said she had to follow the same method as before.

She won the prize again!

Show how Penny did this.

Extension

Another player picked the same four numbers as Penny. This player wrote them as shown at the right. Explain why he won the grand prize.

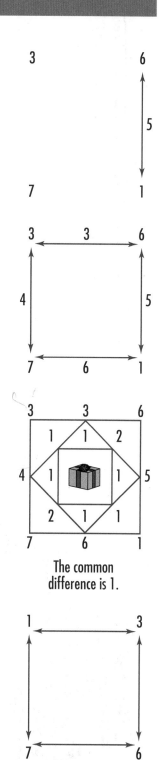

The common difference is 1.

CUMULATIVE REVIEW

Copy and complete.

1. 30 = ■ ones
2. 50 = ■ tens
3. 900 = ■ hundreds
4. 4,200 = ■ hundreds
5. 180 = ■ tens
6. 3,230 = ■ tens

Compare using <, >, or =.

7. 43 ● 48
8. 670 ● 660
9. 742 ● 745
10. 396 ● 396
11. 2,303 ● 2,330
12. 8,854 ● 8,834

Round to the nearest hundred.

13. 212
14. 346
15. 450
16. 371
17. 2,539
18. 4,281

Copy and complete.

19. 152 + ■ = 152
20. 8 + 6 = 6 + ■
21. (5 + 4) + 9 = 5 + (■ + 9)

Find the missing addend.

22. 9 + ■ = 12
23. ■ + 5 = 11
24. 5 + ■ = 12
25. 7 + ■ = 13

Simplify.

26. (5 + 6) − 4
27. (15 − 9) + 7
28. 7 + (18 − 9)

Estimate the sum or difference. Answers are based on rounding.

29. 42
    + 23

30. 76
    − 38

31. 392
    − 228

32. 576
    + 322

33. 3,249
    + 5,894

Choose the most reasonable answer.

34. Jane has 35 soccer balls. Less than 10 of them are yellow. How many soccer balls are yellow?

    a. 35
    b. 16
    c. 11
    d. 7

35. Jim loves to read. He has 350 books in his room. Some of the books are fiction, and more than 200 are nonfiction. How many books are fiction?

    a. 200
    b. 75
    c. 300
    d. 250

2.7 Mental Math: Subtraction Facts

Objective: to use mental math strategies for subtraction

Lee buys 27 pieces of candy at the candy store. He gives some of the candy to his sister, Katherine. Now he has 15 pieces left. How many pieces does Lee give to Katherine?

Write a subtraction fact to model the problem.
27 – ■ = 15

Look at the place values. 27 has one more ten and two more ones than 15. One ten and two ones is 12. What number subtracted from 27 gives you 15?
27 – 12 = 15

Check your answer by adding.
15 + 12 = 27

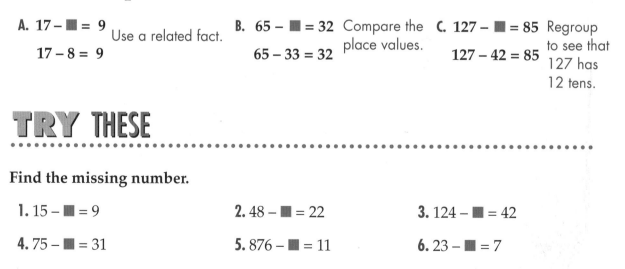

Lee gives Katherine 12 pieces of candy.

More Examples

A. 17 – ■ = 9   Use a related fact.
17 – 8 = 9

B. 65 – ■ = 32   Compare the place values.
65 – 33 = 32

C. 127 – ■ = 85   Regroup to see that 127 has 12 tens.
127 – 42 = 85

TRY THESE

Find the missing number.

1. 15 – ■ = 9

2. 48 – ■ = 22

3. 124 – ■ = 42

4. 75 – ■ = 31

5. 876 – ■ = 11

6. 23 – ■ = 7

Exercises

Find the missing number.

1. $16 - \blacksquare = 9$

2. $67 - \blacksquare = 60$

3. $129 - \blacksquare = 90$

4. $346 - \blacksquare = 100$

5. $49 - \blacksquare = 7$

6. $759 - \blacksquare = 654$

7. $1{,}478 - \blacksquare = 478$

8. $4{,}899 - \blacksquare = 799$

★**9.** $34{,}580 - \blacksquare = 30{,}410$

PROBLEM SOLVING

10. Caroline made 36 cookies. She gave some to her friends. Now she has 3 cookies left. How many cookies did she give away?

11. The soccer team had a total of 27 points. Conner made some of the goals and his teammates made 19. How many goals did Conner make?

MIND BUILDER

Mental Math Strategies

Add 9 to each number. (You can also add 10 and subtract 1.)
Example: $78 + 10 = 88 \, (-1) = 87$

1. 52 **2.** 45 **3.** 41 **4.** 18 **5.** 167 **6.** 84

Subtract 9 from each number. (You can also subtract 10 and add 1.)
Example: $78 - 10 = 68 \, (+1) = 69$

7. 27 **8.** 84 **9.** 97 **10.** 76 **11.** 144 **12.** 25

Add 19 to each number. (You can also add 20 and subtract 1.)
Example: $78 + 20 = 98 \, (-1) = 97$

13. 64 **14.** 126 **15.** 32 **16.** 59 **17.** 71 **18.** 86

Subtract 19 from each number. (You can also subtract 20 and add 1.)
Example: $78 - 20 = 58 \, (+1) = 59$

19. 58 **20.** 73 **21.** 65 **22.** 79 **23.** 56 **24.** 120

2.8 Subtracting Greater Numbers

Objective: to subtract greater numbers

The Chen family saw the dolphin show at Oceanland Park. How much more does the dolphin weigh than the seal?

Weights of Animals at Oceanland Park (in pounds)	
Dolphin	912
Seal	197
Shark	1,099
Whale	8,178

THINK
An estimate is $900 - 200 = 700$.

Step 1	Step 2	Step 3
Subtract the ones.	Subtract the tens.	Subtract the hundreds.
$\begin{array}{r} {}^{0\ 12}\!\!\!\!\!\!\!\! \\ 9\cancel{1}2 \\ -197 \\ \hline 5 \end{array}$	$\begin{array}{r} {}^{8\ 10\ 12}\!\!\!\!\!\!\!\! \\ \cancel{9}\cancel{1}2 \\ -197 \\ \hline 15 \end{array}$	$\begin{array}{r} {}^{8\ 10\ 12}\!\!\!\!\!\!\!\! \\ \cancel{9}\cancel{1}2 \\ -197 \\ \hline 715 \end{array}$

The dolphin weighs 715 more pounds than the seal.
Is the answer reasonable?

The Chen family also learned about whales and sharks. How much more does the whale weigh than the shark?

THINK
An estimate is $8,000 - 1,000 = 7,000$.

Step 1	Step 2	Step 3	Step 4
Subtract the ones.	Subtract the tens.	Subtract the hundreds.	Subtract the thousands.
$\begin{array}{r} {}^{6\ 18}\!\!\!\!\!\! \\ 8,17\cancel{8} \\ -1,099 \\ \hline 9 \end{array}$	$\begin{array}{r} {}^{0\ 16\ 18}\!\!\!\!\!\! \\ 8,\cancel{1}7\cancel{8} \\ -1,099 \\ \hline 79 \end{array}$	$\begin{array}{r} {}^{0\ 16\ 18}\!\!\!\!\!\! \\ 8,\cancel{1}7\cancel{8} \\ -1,099 \\ \hline 079 \end{array}$	$\begin{array}{r} {}^{0\ 16\ 18}\!\!\!\!\!\! \\ 8,\cancel{1}7\cancel{8} \\ -1,099 \\ \hline 7,079 \end{array}$

The whale weighs 7,079 more pounds than the shark. Is the answer reasonable?

TRY THESE

Subtract.

1. 8,421
  − 3,789

2. $62.01
  − 17.53

3. 26,978
  − 17,889

4. 34,071
  − 25,760

Exercises

Subtract.

1. $97,624
  − 8,976

2. 21,407
  − 6,824

3. 72,578
  − 21,936

4. 48,803
  − 21,738

5. 67,453
  − 8,549

6. $502.88
  − 73.89

7. 92,352
  − 3,508

8. 82,034
  − 63,666

9. $46,985
  − 37,998

10. $46,947
  − 495

11. $876,271
  − 436,865

12. $5,234.60
  − 4,372.51

Simplify.

13. 2,800 − (1,658 + 1,079)

14. (2,800 − 1,658) + 1,079

PROBLEM SOLVING

15. In the 1990 census, Rochester, NY, reported a population of 230,356 and Seneca Falls reported a population of 7,370. How many more people lived in Rochester than in Seneca Falls?

16. The Georgia Dome holds 71,280 people. The Ericsson Stadium in North Carolina holds 72,520 people. How many more people does the Ericsson Stadium hold than the Georgia Dome?

17. A dolphin eats 231 pounds of food a week. A whale eats 805 pounds of food a week. How much more food does a whale eat each week than a dolphin?

★**18.** Copy and complete. Subtract across. Then subtract down.

8,142	2,375	?
5,086	1,978	?
?	?	?

19. Find the number for each shape.

● − ■ = 9

● + ● = 18

2.9 Problem-Solving Application: Making Change

Objective: to solve problems involving making correct change

While at the zoo, Vicki buys an ice cream cone that costs $1.67. She gives $2.00 to the clerk. How much change should the clerk give Vicki?

Use the four-step plan to solve the problem.

 1. READ
You know how much the ice cream cone costs, and you know how much Vicki gave the clerk. You need to know what change the clerk should give Vicki.

 2. PLAN
Start with the cost of the ice cream cone and count up to the amount given.

3. SOLVE

$1.67 ⟶ $1.68 → $1.69 → $1.70 → $1.75 ⟶ $2.00

3 pennies, 1 nickel, and 1 quarter add up to 33¢. The clerk should give Vicki 33¢ in change.

 4. CHECK
Add to check the answer.

$$\begin{array}{r} \overset{1\ \ 1}{\$1.67} \\ +\ 0.33 \\ \hline \$2.00 \end{array} \checkmark$$

33¢ = $0.33
The answer checks.

TRY THESE

Solve. List the coins and bills needed for change. Try to use as few as possible. The coins of least value should be written first.

1. Gino buys an ice cream cone that costs 32¢. He gives the clerk 50¢. How much change should the clerk give Gino?

2. Mrs. Cohen buys a sun visor that costs $2.50. She gives the clerk $5.00. How much change should the clerk give Mrs. Cohen?

Solve . . . .

Use the prices shown to solve.

1. Skip bought a cookie and lemonade. How much did he spend?

2. Yasmin bought a lemonade. She gave the clerk $2.00. How much change should the clerk have given Yasmin?

3. Molly has $5.00. Does she have enough to buy a slice of pizza and lemonade?

4. Laura gave the clerk $5.00 for a hot dog and lemonade. The clerk gave her 3 quarters in change. Is this the correct change?

5. Greg gave the clerk $1.00 to purchase one item. He received 2 dimes and 2 pennies in change. What did Greg buy?

PRICES

Ice Cream	$1.15
Lemonade	$1.67
Cookie	$0.99
Fruit	$0.78
Hot Dog	$2.46
Pizza	$3.45

6. Erin handed the clerk $20.00 to pay for some food. She received 1 dollar, 2 quarters, 2 dimes, and 4 pennies in change. How much did the food cost?

Make a list of at least *three* possible combinations of coins and bills to make change for the following problems. Do not use more than 4 pennies in one combination. Circle the combination with the least number of coins.

	Price	Amount Given
7.	26¢	50¢
8.	$3.60	$6.40
9.	$4.50	$9.52
10.	$15.05	$20.00

CONSTRUCTED RESPONSE

Use the prices on the chart in the picture above to solve.

11. Andrew buys ice cream for himself and 3 friends. Is $5.00 enough money? Explain how you know that your answer is correct.

CHAPTER 2 REVIEW

LANGUAGE and CONCEPTS

Choose the correct word to complete each sentence.

1. You can use (addition, subtraction) to find the total amount.

2. You can use (addition, subtraction) to find how many more are in one group than in another.

3. The answer in an addition problem is called the (difference, sum).

4. The answer in a subtraction problem is called the (difference, sum).

5. 8 + 0 = 8 shows the (Commutative, Identity) Property of Addition.

6. 9 + 6 = 6 + 9 shows the (Associative, Commutative) Property of Addition.

7. (3 + 4) + 0 = 3 + (4 + 0) shows the (Associative, Identity) Property.

SKILLS and PROBLEM SOLVING

Translate each of the following. Tell if it is an expression or an equation. (Section 2.2)

8. 12 divided by 2

9. 9 times 1

10. 13 decreased by 11 is 2.

11. 19 added to 21 is equal to 40.

12. Seven multiplied by four is twenty-eight.

13. The difference of 91 and 37 is 54.

14. the sum of 1998 and 2008

15. the quotient of 5 and 25

16. 9 more than 15 is 24.

17. 64 divided by 8

18. the total of 3,100 and 8,722

19. Fourteen subtract 10 is four.

Simplify. (Section 2.3)

20. $14 - (3 + 4)$

21. $(15 - 6) + 4$

22. $9 + (13 - 7)$

23. $(30 + 40) - 50$

24. $90 - (20 + 60)$

25. $70 + (60 - 30)$

Estimate by rounding. (Section 2.4)

26.
$$\begin{array}{r} 23 \\ + 58 \end{array}$$

27.
$$\begin{array}{r} 84 \\ - 42 \end{array}$$

28.
$$\begin{array}{r} 2,157 \\ + 436 \end{array}$$

29.
$$\begin{array}{r} 1,892 \\ - 275 \end{array}$$

30.
$$\begin{array}{r} 3,204 \\ + 6,338 \end{array}$$

Add. (Section 2.5)

31. $7.09
     + 2.17

32. 462
     + 54

33. $523
     + 296

34. 198
     + 175

35. $3.57
     + 4.59

36. 787
     + 661

37. 6,521
     + 2,734

38. $785.64
     + 467.96

39. 736 + 492 + 988

40. $498 + $648 + $536

Subtract. (Sections 2.7–2.9)

41. 70
     − 57

42. $85
     − 29

43. 4,745
     − 159

44. $6.20
     − 5.97

45. $700
     − 351

46. 1,804
     − 297

47. $34.46
     − 0.72

48. 58,005
     − 39,243

Choose the most reasonable answer. (Section 2.6)

49. Maria saved $34. Jean saved more than Maria. How much did Jean save?

a. $27   **b.** $34   **c.** $37

Solve. (Section 2.9)

50. Dale had $1.00. He bought a subway ticket that cost 70¢. Now he has two coins in his pocket. What are they?

51. Marie buys peanuts for 45¢ to feed to the elephants. She gives the clerk $1.00. How much change should the clerk give Marie?

52. Mr. Cohen buys a zoo program that costs $1.00. He gives the clerk $10.00. How much change should Mr. Cohen receive?

53. Dave buys a bicycle for $567. He gives the clerk $600. What is his change?

CHAPTER 2 TEST

Simplify.

1. $(5 + 2) - 4$ **2.** $14 - (6 + 6)$ **3.** $(15 - 9) + 5$

4. $12 - (3 + 2)$ **5.** $(8 + 9) - 7$ **6.** $(50 - 20) + 40$

Estimate by rounding.

7. $\begin{array}{r} 52 \\ + 44 \\ \hline \end{array}$ **8.** $\begin{array}{r} 832 \\ + 411 \\ \hline \end{array}$ **9.** $\begin{array}{r} 656 \\ - 319 \\ \hline \end{array}$ **10.** $\begin{array}{r} 7,092 \\ - 6,110 \\ \hline \end{array}$

Add.

11. $\begin{array}{r} 24 \\ + 56 \\ \hline \end{array}$ **12.** $\begin{array}{r} 63 \\ + 17 \\ \hline \end{array}$ **13.** $\begin{array}{r} \$19 \\ + 87 \\ \hline \end{array}$ **14.** $\begin{array}{r} \$3.94 \\ + 5.29 \\ \hline \end{array}$

15. $\begin{array}{r} 864 \\ + 338 \\ \hline \end{array}$ **16.** $\begin{array}{r} 339 \\ 97 \\ + 553 \\ \hline \end{array}$ **17.** $\begin{array}{r} \$4,361 \\ + 1,724 \\ \hline \end{array}$ **18.** $\begin{array}{r} 3,756 \\ + 6,624 \\ \hline \end{array}$

19. $\$7.97 + \$5.37$ **20.** $245 + 28 + 61$ **21.** $5,689 + 3,788$

Subtract.

22. $\begin{array}{r} 61 \\ - 26 \\ \hline \end{array}$ **23.** $\begin{array}{r} 50 \\ - 11 \\ \hline \end{array}$ **24.** $\begin{array}{r} 747 \\ - 359 \\ \hline \end{array}$ **25.** $\begin{array}{r} 845 \\ - 57 \\ \hline \end{array}$

26. $\begin{array}{r} \$2.31 \\ - 1.64 \\ \hline \end{array}$ **27.** $\begin{array}{r} 602 \\ - 457 \\ \hline \end{array}$ **28.** $\begin{array}{r} 4,000 \\ - 2,678 \\ \hline \end{array}$ **29.** $\begin{array}{r} 28,135 \\ - 16,409 \\ \hline \end{array}$

30. $4,233 - 1,954$ **31.** $\$6,175 - \$1,286$ **32.** $60,209 - 4,967$

Solve.

33. The theater sold a total of 8,024 tickets. It sold 189 adult tickets. The remaining tickets were children's tickets. How many children's tickets were sold?

34. Mrs. Chang spends $6.25 for zoo tickets. She gives the clerk $10.00. How much change does she receive?

Magic Squares

Magic squares have been in many different cultures on Earth for over 4,000 years. They have been found carved on stone, engraved on jewelry, or used as art. The most commonly recognized type is the 3 × 3 grid, but they also can be found in other grid sizes.

19	12	17
14	16	18
15	20	13

In a regular 3 × 3 grid, a magic square has the following characteristics:

- The numbers are all consecutive, or follow each other in order. For example, the numbers 16, 17, 18, 19, 20, 21, 22, 23, and 24 are consecutive.
- The numbers add up to the same sum whether you add them horizontally, vertically, or diagonally. The preceding magic square adds up to 48 in all directions.

In some 4 × 4 magic squares, not only are the numbers consecutive and each row adds up to the same number, but so do the four outer corners. The four inside squares in the center also add up to the magic sum.

16	3	2	13
5	10	11	8
9	6	7	12
4	15	14	1

Test the magic square to the right.

1. Find the sum of each of the rows going across.

2. Find the sum of each of the rows going down.

3. Find the sum of the two diagonals.

4. Find the sum of the four corner numbers.

5. Find the sum of the four inside numbers: 10 + 11 + 6 + 7.

6. What is the magic sum?

Copy and complete the following 4 × 4 magic squares.

7.

19	5	■	16
8	■	13	11
12	10	9	■
■	17	18	4

8.

25	11	12	■
■	20	19	17
18	■	15	21
13	23	24	■

CUMULATIVE TEST

1. Name the property.
1,706 + 0 = 1,706
- **a.** Associative Property
- **b.** Commutative Property
- **c.** Identity Property
- **d.** none of the above

2. Simplify.
(317 − 309) + 6
- **a.** 9
- **b.** 10
- **c.** 12
- **d.** none of the above

3. Estimate by rounding.
2,756
+ 4,872
- **a.** 6,000
- **b.** 7,000
- **c.** 8,000
- **d.** none of the above

4.
78,361
43,512
+ 26,127
- **a.** 20,000
- **b.** 30,000
- **c.** 140,000
- **d.** none of the above

5.
5,349
− 2,876
- **a.** 2,363
- **b.** 2,473
- **c.** 3,463
- **d.** none of the above

6. Which number is one thousand, three hundred forty-six?
- **a.** 1,257
- **b.** 1,346
- **c.** 1,357
- **d.** none of the above

7.
304,942
− 218,337
- **a.** 84,515
- **b.** 86,605
- **c.** 114,279
- **d.** none of the above

8. 72,093 − 5,496 = _____
- **a.** 66,597
- **b.** 66,617
- **c.** 73,403
- **d.** none of the above

9. Jenny bought a headband and a wristband. How much is Jenny's change?

WRISTBAND $1.39
HEADBAND $1.85

- **a.** $1.76
- **b.** $2.76
- **c.** $3.24
- **d.** none of the above

10. Use the chart. How many more quarters than nickels are made per month?

U.S. Coin Production Per Month	
Pennies	854,783,700
Nickels	110,306,250
Dimes	194,608,330
Quarters	155,616,660
Half-dollars	2,559,166

- **a.** 46,310,410
- **b.** 45,410,310
- **c.** 45,310,410
- **d.** none of the above

CHAPTER 3

Multiplying by Whole Numbers

Lisa Binner
Missouri

3.1 The Meaning of Multiplication

Objective: to identify multiplication as repeated addition

Rachel puts her scallop shells in 3 rows. There are 4 shells in each row. To find the total number of shells, she can add or multiply.

When there are equal groups, you can multiply to find how many in all.

There are 12 shells in all.

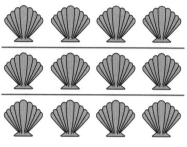

Add: $4 + 4 + 4 = 12$ 3 fours

Multiply: $3 \times 4 = 12$
$\times$ is read as *times*.

More Examples

You can use pictures or number lines to show multiplication.

A. $4 \times 2 = \blacksquare$
Draw 4 groups of 2.

$2 + 2 + 2 + 2 = 8$
$4 \times 2 = 8$

B. $2 \times 5 = \blacksquare$
Draw 2 jumps of 5.

Start at zero.

$5 + 5 = 10$
$2 \times 5 = 10$

C. $3 \times 6 = \blacksquare$
Draw 3 rows with 6 in each row.

$$\begin{array}{r} 6 \\ 6 \\ +\ 6 \\ \hline 18 \end{array}$$
$3 \times 6 = 18$

D. $4 \times 8 = \blacksquare$
Draw a 4 by 8 rectangle.

$$\begin{array}{r} 8 \\ 8 \\ 8 \\ +\ 8 \\ \hline 32 \end{array}$$
$4 \times 8 = 32$

TRY THESE

Match the picture to the multiplication fact it shows.

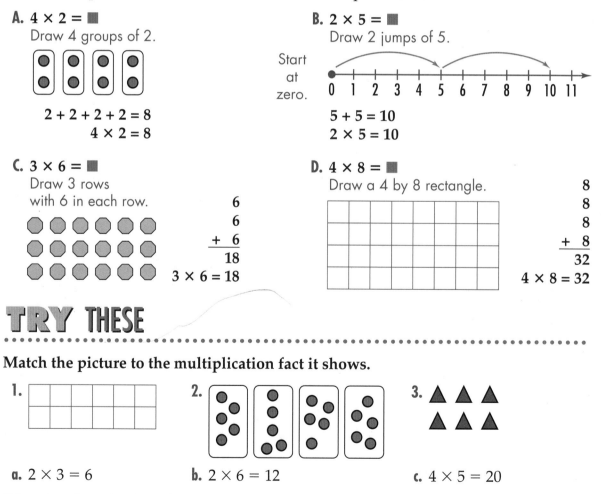

1.

2.

3.

a. $2 \times 3 = 6$

b. $2 \times 6 = 12$

c. $4 \times 5 = 20$

Exercises

Add. Then multiply.

1. 7 + 7 + 7
3 × 7

2. 8 + 8 + 8
3 × 8

3. 4 + 4
2 × 4

4. 9 + 9
2 × 9

5. 6 + 6 + 6
3 × 6

6. 2 + 2
2 × 2

7. 5 + 5
2 × 5

8. 9 + 9 + 9
3 × 9

9. 7 + 7
2 × 7

Draw a picture to show how many. Then add or multiply.

10. 5 rows of 6
5 × 6 = ■

11. 6 rows of 1
6 × 1 = ■

12. 4 rows of 7
4 × 7 = ■

PROBLEM SOLVING

Write the multiplication fact. Solve.

13. There are 7 days in a week. How many days are there in 8 weeks?

14. The recital room has 11 rows of chairs. There are 6 chairs per row. How many chairs are in the recital room?

15. The class was divided into four equal groups of four. How many students were in the class?

16. The Howard family reserved ten hotel rooms per hotel floor for the reunion. The hotel has four floors. How many rooms did the Howard family reserve for their reunion?

TEST PREP

Choose the multiplication fact that solves each problem.

17. There are 4 rows of buckets with 6 buckets in each row. How many buckets are there?

a. 6 × 3 = 18

b. 4 × 6 = 24

18. There are 3 buckets with 7 shells in each bucket. How many shells are there?

a. 3 × 7 = 21

b. 3 × 6 = 18

3.2 Properties of Multiplication

Objective: to identify and use the properties of multiplication

Sophia ordered some sheets of plant stickers from a catalog. To find the total cost of each kind, she uses multiplication facts.

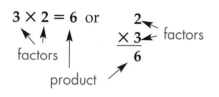

3 × 2 = 6 or

factors

product

Say: 3 times 2 is equal to 6.

Properties of multiplication can help you remember basic facts.

Stickers-by-Mail			
Quantity	**Item**	**Cost Each**	**Total Cost**
3	Daisies	$2	$6
1	Red Roses	$8	$8
2	Ferns	$3	$6
2	Tulips	0	0

FREE
Two sheets of tulip stickers with purchase of $10.00 or more

▶ **COMMUTATIVE PROPERTY OF MULTIPLICATION**
Changing the order of the factors does not change the **product**.

3	Daisies	$2	$6
2	Ferns	$3	$6

3 × 2 = 6   2 × 3 = 6

▶ **IDENTITY PROPERTY OF MULTIPLICATION**
When one factor is 1, the product is the same as the other factor.

1	Red Roses	$8	$8

1 × 8 = 8   8 × 1 = 8

▶ **ZERO PROPERTY OF MULTIPLICATION**
When one factor is 0, the product is 0.

2	Tulips	0	0

2 × 0 = 0   0 × 2 = 0

▶ **ASSOCIATIVE PROPERTY OF MULTIPLICATION**
The way factors are grouped does not change the product.

$(3 \times 2) \times 5 \overset{?}{=} 3 \times (2 \times 5)$

$6 \quad \times 5 \overset{?}{=} 3 \times \quad 10$

$30 = 30$

You can use the *Commutative Property* and the *Associative Property* of multiplication to find the product. Look for factors that have 10 or a multiple of 10 as a product.

More Examples

A. $5 \times 7 \times 2 \longrightarrow 7 \times 5 \times 2 \longrightarrow 7 \times (5 \times 2) \longrightarrow 7 \times 10 \longrightarrow 70$

Change the order of the factors 5 and 7.   Group the factors 5 and 2. $5 \times 2 = 10$

Change to factors that are basic facts.

B. $4 \times 18 \longrightarrow 4 \times (2 \times 9) \longrightarrow (4 \times 2) \times 9 \longrightarrow 8 \times 9 \longrightarrow 72$

C. $3 \times 14 \longrightarrow 3 \times (2 \times 7) \longrightarrow (3 \times 2) \times 7 \longrightarrow 6 \times 7 \longrightarrow 42$

TRY THESE

Name the property.

1. $25 \times 0 = 0$

2. $4 \times 8 = 8 \times 4$

3. $(5 \times 2) \times 3 = 5 \times (2 \times 3)$

4. $9,872 \times 1 = 9,872$

5. $13 \times 9 = 9 \times 13$

Exercises

Multiply without using paper and pencil.

1. $5 \times 9 \times 2$ **2.** $8 \times 7 \times 5$ **3.** 5×24 **4.** 3×27

5. 7×16 **6.** 2×15 **7.** 3×24 **8.** 3×15

9. 18×3 **10.** 4×25 **11.** 50×4 **12.** 4×16

13. 5×36 **14.** 8×35 **15.** 16×5 **16.** 24×5

PROBLEM SOLVING

17. If the product of two numbers is zero, what must be true? Which property helps you to solve this problem?

★ 18. One row of bleachers can hold 124 students. How many students will 5 rows hold?

3.3 Distributive Property

Objective: to identify and use the Distributive Property

When Sophia places her collection of red stickers next to her collection of blue stickers, there are five rows of seven stickers (5 × 7).

There are 5 × 3 or 15 red stickers.
There are 5 × 4 or 20 blue stickers.
Altogether there are (5 × 3) + (5 × 4) or 35 stickers.

$$5 \times 7 = 5 \times (3 + 4)$$
$$= (5 \times 3) + (5 \times 4)$$
$$= 15 + 20$$
$$= 35$$

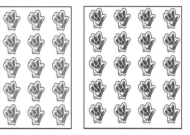

This illustrates the **Distributive Property**. The Distributive Property allows you to multiply a sum by multiplying each addend separately, and then adding their products.

The Distributive Property helps you to multiply larger factors. You begin by replacing the larger factor with an addition fact that equals the factor.

More Examples

A. $4 \times 17 = 4 \times (10 + 7)$
$$= (4 \times 10) + (4 \times 7)$$
$$= 40 + 28$$
$$= 68$$

B. $3 \times 29 = 3 \times (20 + 9)$
$$= (3 \times 20) + (3 \times 9)$$
$$= 60 + 27$$
$$= 87$$

TRY THESE

Find the missing numbers.

1. $7 \times 13 = 7 \times (\blacksquare + \blacksquare)$
$$= (\blacksquare \times 10) + (7 \times \blacksquare)$$
$$= 70 + \blacksquare$$
$$= \blacksquare$$

2. $6 \times 28 = \blacksquare \times (20 + \blacksquare)$
$$= (6 \times \blacksquare) + (6 \times \blacksquare)$$
$$= \blacksquare + 48$$
$$= \blacksquare$$

Exercises

Solve using the Distributive Property.

1. $4 \times 21 = (4 \times 20) + (4 \times 1)$

2. $7 \times 14 = (7 \times 10) + (7 \times 4)$

3. 12×5

4. 80×6

5. 50×4

6. 6×14

7. 5×15

8. 9×17

PROBLEM SOLVING

9. Mrs. Matino bought 15 packages of markers. There were 9 markers in each package. How many markers did she buy?

10. The teacher wants each student to have 12 pencils. There are 5 students. How many pencils must the teacher buy?

CONSTRUCTED RESPONSE

11. Isabella has 18 sticker pages. Each page has 9 stickers on it. How many stickers does Isabella have? Explain your answer.

12. One factor is 9. The product is 9. What is the other factor? Explain your reasoning.

MIXED REVIEW

Name the property.

13. $(8 \times 7) \times 5 = 8 \times (7 \times 5)$

14. $87 \times 0 = 0$

15. $13 \times 19 = 19 \times 13$

16. $3{,}421 \times 1 = 3{,}421$

17. $4 \times (9 + 2) = (4 \times 9) + (4 \times 2)$

18. $3 \times (4 \times 9) = (4 \times 9) \times 3$

3.4 Factors and Multiples

Objective: to understand the difference between factors and multiples; to generate multiples and factors of a number

Houses on York Street are numbered 2, 4, 6, 8, 10, 12, and 14. These house numbers are multiples of 2.

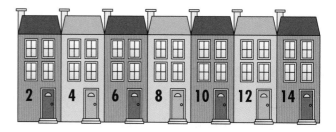

A **multiple** is a number found by multiplying a given number by a whole number.

Juan lives at 14 York Street. If you multiply two numbers together to create 14, those numbers are called factors. The factors of 14 are 1, 2, 7, and 14.

A **factor** is one of two or more numbers that when multiplied together produce a given product.

More Examples

A. The multiples of 6 are 0, 6, 12, 18, 24, and so on.

$$0 \times 6 = 0$$
$$1 \times 6 = 6$$
$$2 \times 6 = 12$$
$$3 \times 6 = 18$$
$$4 \times 6 = 24$$

B. The factors of 6 are 1, 2, 3, and 6.

$$1 \times 6 = 6$$
$$2 \times 3 = 6$$
$$3 \times 2 = 6$$
$$6 \times 1 = 6$$

TRY THESE

Find the products to complete the following tables.

1.

×	1	2	3	4	5	6	7	8	9
4	4	8	12	■	■	■	■	32	■

2.

×	1	2	3	4	5	6	7	8	9
9	9	18	■	■	■	■	63	■	■

3.

×	1	2	3	4	5	6	7	8	9
6	6	12	■	■	■	■	42	■	■

4.

×	1	2	3	4	5	6	7	8	9
7	7	14	■	■	■	■	■	■	63

Solve.

5. Find the factors of 5.

6. Find the factors of 10.

Exercises

Complete the pattern of multiples.

1. 0, 4, 8, 12, ■, ■, ■, ■, ■

2. 0, 7, 14, 21, ■, ■, ■, ■, ■

3. 0, 3, 6, 9, 12, 15, ■, ■, ■, ■, ■

4. ■, ■, 16, 24, ■, ■, 48, ■

Write *true* or *false*. If the statement is false, rewrite it to make it true.

5. 7 is a multiple of 2.

6. 20 is a factor of 2.

7. 2 is a factor of 20.

8. 21 is a multiple of 7.

9. 23 is a multiple of 3.

10. 4 is a factor of 24.

Find all of the factors for the following numbers.

11. 8   **12.** 12   **13.** 15   **14.** 20

PROBLEM SOLVING

15. List the first ten multiples of 2. List the first ten multiples of 5. What numbers are common to both lists?

16. There are 5 streets with 9 houses on each street. How many houses are there in the neighborhood?

17. Jennifer lines up her freshly baked sugar cookies. She has 8 rows of 9 cookies. How many cookies did she bake?

CONSTRUCTED RESPONSE

18. List the multiples of your age. Then list the factors of your age. Write about what you notice.

3.5 Multiplication Patterns

Objective: to locate patterns to multiply mentally

In baseball for youths, the bases are 60 feet apart. To find how far a player runs after hitting a home run, multiply 60 by 4.

Say: 4 times 6 tens is 24 tens.

Write: $4 \times 60 = 240$

The distance around the bases is 240 feet. The distance between the bases also could be added as $60 + 60 + 60 + 60 = 240$ to find the total distance.

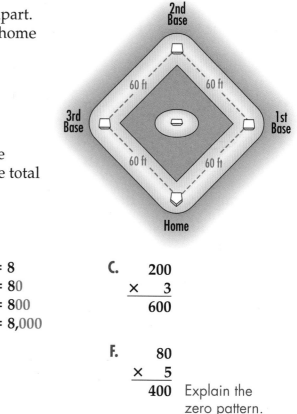

More Examples

A.
$7 \times 1 = 7$
$7 \times 10 = 70$
$7 \times 100 = 700$
$7 \times 1{,}000 = 7{,}000$

B.
$2 \times 4 = 8$
$2 \times 40 = 80$
$2 \times 400 = 800$
$2 \times 4{,}000 = 8{,}000$

C.
$$\begin{array}{r} 200 \\ \times 3 \\ \hline 600 \end{array}$$

D.
$$\begin{array}{r} 500 \\ \times 7 \\ \hline 3{,}500 \end{array}$$
What pattern with zeros do you see?

E.
$$\begin{array}{r} 4{,}000 \\ \times 6 \\ \hline 24{,}000 \end{array}$$

F.
$$\begin{array}{r} 80 \\ \times 5 \\ \hline 400 \end{array}$$
Explain the zero pattern.

There is at least the same number of zeros in the product as in the two factors.

TRY THESE

Say and write the products.

1. 3×1
3×10
3×100
$3 \times 1{,}000$

2. 2×9
2×90
2×900
$2 \times 9{,}000$

Exercises

Multiply.

1. a. 7×6 **b.** 7×60 **c.** 7×600 **d.** $7 \times 6,000$

2. a. 5×4 **b.** 5×40 **c.** 5×400 **d.** $5 \times 4,000$

3.
$$\begin{array}{r} 10 \\ \times\ 5 \\ \hline \end{array}$$

4.
$$\begin{array}{r} 100 \\ \times\ 2 \\ \hline \end{array}$$

5.
$$\begin{array}{r} 1,000 \\ \times\ 4 \\ \hline \end{array}$$

6.
$$\begin{array}{r} 400 \\ \times\ 9 \\ \hline \end{array}$$

7.
$$\begin{array}{r} 90 \\ \times\ 6 \\ \hline \end{array}$$

Find the missing factor.

8. $6 \times \blacksquare = 600$

9. $2 \times \blacksquare = 8,000$

10. $7 \times \blacksquare = 490$

11. $\blacksquare \times 100 = 800$

12. $\blacksquare \times 700 = 3,500$

★ **13.** $\blacksquare \times 2,000 = 10,000$

PROBLEM SOLVING

14. In professional baseball, the bases are 90 feet apart. If a player hits a triple, how far will the player run to reach third base?

15. By the eighth inning, 7 people had left the game. At the start of the game, there were 50 people. How many people were still at the game?

16. There are 20 baseball cards in 1 package. Find the number of cards in 8 packages.

17. The product of two numbers is 24. The sum is 11. What are the two numbers?

18. There are 80 students in each grade at Hillcrest School. There are 6 grades. How many students go to Hillcrest School?

19. If a baseball team hits 300 home runs a season, how many home runs would they hit in 6 seasons if they followed the same trend?

3.6 Problem-Solving Strategy: Making an Organized List

Objective: to solve problems by creating an organized list

Some problems are easy to solve if you make a list. Study the following example.

A local athletic club serves four kinds of juice. Each comes in a small size or a large size. How many different juice orders are possible?

1. READ How many juice orders are possible? You know that there are four kinds of juice, and each comes in two sizes.

2. PLAN List the different juice orders. Then count the number of different orders.

3. SOLVE

small orange	*small apple*
large orange	*large apple*
small grape	*small tomato*
large grape	*large tomato*

There are eight different juice orders.

When you make a list, try to organize the information.

4. CHECK There are four kinds of drinks and two sizes, so eight possible juice orders makes sense.

TRY THESE

How many different ways are possible? Make a list.

1. peach or banana; and milk or juice

2. carrots, peas, or beans; and ham, turkey, or beef

Solve......

Use any strategy.

1. Sharon has a blue skirt and a brown skirt. She has five blouses that she can wear with either skirt. The blouses are white, tan, yellow, blue, and pink. How many different outfits can she wear?

2. The athletic club sells many items in the club store. Tennis balls cost $2. Sweatbands cost $1. If you buy at least one of each, in how many ways can you spend exactly $8?

3. Use the ballot. How many different ways can a person vote?

4. Use the menu. How many different lunches are possible?

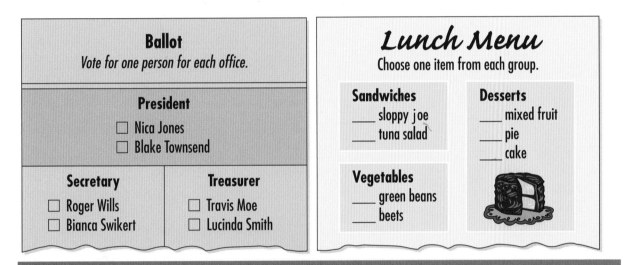

Ballot
Vote for one person for each office.

President
☐ Nica Jones
☐ Blake Townsend

Secretary
☐ Roger Wills
☐ Bianca Swikert

Treasurer
☐ Travis Moe
☐ Lucinda Smith

Lunch Menu
Choose one item from each group.

Sandwiches
____ sloppy joe
____ tuna salad

Desserts
____ mixed fruit
____ pie
____ cake

Vegetables
____ green beans
____ beets

MID-CHAPTER REVIEW

Multiply mentally.

1. 5×300
2. $7 \times 1,000$
3. 6×50
4. 6×11

Estimate.

5. 2×45
6. $9 \times 4 \times 5$
7. 33×8
8. 179×3

Solve.

9. There are 13 muffins in a baker's dozen. How many muffins are in 6 baker's dozens?

Problem Solving

Logic Problems

Copy and complete the grids to solve each problem using the given clues. Mark each box X for no or *yes* for yes.

A. Rosa, Mai Li, and Susan attend the same ballet school, but they travel there in different ways. One rides the bus; one walks; and one is driven to class by her mother. Who rides the bus, who walks, and who is driven to class?

	bus	car	walk
Rosa			
Mai Li			
Susan			

Clues

1. Rosa was late to class last week because of a flat tire.

2. Mai Li and the girl who rides the bus are best friends.

3. There is no public transportation near Susan's home.

4. Mai Li's mother stays home with her new baby sister while she is at class.

B. Vanessa, Maxine, and Yvonne are sisters. Who is the oldest sister? Who is the middle sister? Who is the youngest sister?

	oldest	middle	youngest
Vanessa			
Maxine			
Yvonne			

Clues

1. Vanessa is younger than Maxine.

2. Yvonne is not the youngest.

3. Maxine is older than Yvonne.

CUMULATIVE REVIEW

Round to the nearest hundred.

1. 427　　　　2. 562　　　　3. 309　　　　4. 1,228　　　　5. 34,687

Round to the nearest thousand.

6. 6,248　　　7. 3,567　　　8. 5,962　　　9. 19,999　　　10. 4,327,426

Round to the nearest ten million.

11. 31,789,232　　　　12. 764,516,722　　　　13. 239,012,510

Find each missing addend.

14. $90 + \blacksquare = 108$　　　15. $\blacksquare + 17 = 25$　　　16. $22 + \blacksquare = 50$

Estimate by rounding.

17. $52 + 19$　　　　18. $632 + 174$　　　　19. $2,936 + 3,475$

Find the sum.

20. $\begin{array}{r} \$4.75 \\ +\ 5.72 \\ \hline \end{array}$　　21. $\begin{array}{r} 737 \\ +\ 418 \\ \hline \end{array}$　　22. $\begin{array}{r} 3,057 \\ +\ 9,386 \\ \hline \end{array}$　　23. $\begin{array}{r} 4,239,167 \\ +\ 16,421,029 \\ \hline \end{array}$

24. $17 + 38 + 16$　　　25. $1,426 + 443 + 2,835$　　　26. $\$625 + \$437 + \$389$

Find the difference.

27. $\begin{array}{r} 1,369,815 \\ -\ 727,834 \\ \hline \end{array}$　　28. $\begin{array}{r} \$1,640 \\ -\ 795 \\ \hline \end{array}$　　29. $\begin{array}{r} 702 \\ -\ 356 \\ \hline \end{array}$

30. $\begin{array}{r} \$7,080 \\ -\ 2,345 \\ \hline \end{array}$　　31. $\begin{array}{r} 750,276,128 \\ -\ 34,619,247 \\ \hline \end{array}$

32. $\$9.06 - \$2.29$　　　33. $8,001 - 2,675$　　　34. $432,129,783 - 8,076,497$

Solve.

35. Ned has 820 pennies and 7 dimes. How much money does he have?

36. Sean has 70 dimes and 962 pennies. How much money does he have?

37. Caroline has 90 dimes. How much money does she have?

3.7 Estimating Products Using Rounding

Objective: to use rounding and estimating to multiply mentally

There were 862 people who came out to cheer for the high school basketball team. Each person paid $4. About how much money was collected?

To solve this problem, multiply. Since you do not need an exact answer, estimate.

One way to estimate uses rounding.

- Round each factor to its greatest place-value position.
- Do not round one-digit factors.
- Then multiply.

$$
\begin{array}{ccc}
862 & \rightarrow & 900 \\
\times\ 4 & & \times\ 4 \\
\hline
& & 3,600
\end{array}
$$
862 rounds to 900.
Do not round the 4 because it is a single digit.

About $3,600 was collected.

More Examples

A.
$$
\begin{array}{ccc}
64 & \rightarrow & 60 \\
\times\ 5 & & \times\ 5 \\
\hline
& & 300
\end{array}
$$

B.
$$
\begin{array}{ccc}
479 & \rightarrow & 500 \\
\times\ 8 & & \times\ 8 \\
\hline
& & 4,000
\end{array}
$$

C.
$$
\begin{array}{ccc}
878 & \rightarrow & 900 \\
\times\ 5 & & \times\ 5 \\
\hline
& & 4,500
\end{array}
$$

D.
$$
\begin{array}{ccc}
820 & \rightarrow & 800 \\
\times\ 9 & & \times\ 9 \\
\hline
& & 7,200
\end{array}
$$

E.
$$
\begin{array}{ccccc}
\$7.98 & \rightarrow & \$8.00 & \rightarrow & \$8 \\
\times\ 3 & & \times\ 3 & & \times\ 3 \\
\hline
& & & & \$24
\end{array}
$$

TRY THESE

Estimate each product.

1.
$$
\begin{array}{ccc}
47 & \rightarrow & 50 \\
\times\ 3 & & \times\ 3
\end{array}
$$

2.
$$
\begin{array}{ccc}
237 & \rightarrow & 200 \\
\times\ 2 & & \times\ 2
\end{array}
$$

3.
$$
\begin{array}{ccc}
452 & \rightarrow & 500 \\
\times\ 4 & & \times\ 4
\end{array}
$$

4.	15,609	5.	2,401	6.	250	7.	520
	$\times\ \ \ \ \ 5$		$\times\ \ \ \ \ 8$		$\times\ \ \ 4$		$\times\ \ \ 9$

Exercises

Estimate each product by rounding.

1.	$271	2.	904	3.	$21.50	4.	6,742
	$\times\ \ \ 9$		$\times\ \ \ 6$		$\times\ \ \ \ \ 3$		$\times\ \ \ 2$

5.	$71	6.	4,516	7.	6,110	8.	5×909
	$\times\ \ \ 4$		$\times\ \ \ 2$		$\times\ \ \ 9$		

9. $9 \times \$66.25$ **10.** $693 \times \$8$

Use estimation to decide which product is reasonable.

11. $3 \times 24 =$	**a.** 72	**b.** 612	**c.** 36,002
12. $8,007 \times 3 =$	**a.** 261	**b.** 2,421	**c.** 24,201
13. $7 \times 181 =$	**a.** 560	**b.** 1,267	**c.** 10,267
14. $0 \times 20,773 =$	**a.** 0	**b.** 20,773	**c.** 2,070,730
15. $89 \times 3 =$	**a.** 762	**b.** 267	**c.** 672
16. $7 \times 8,093 =$	**a.** 566	**b.** 5,661	**c.** 56,651

PROBLEM SOLVING

17. There are 92 students in each grade at Batavia Elementary School. There are 6 grades. About how many students go to Batavia Elementary School?

18. Mrs. Hutson earns $433 a week at the book-binding company. About how much does she make in 5 weeks?

Use the chart to solve.

19. The chart at the right shows how many shots Monica made and how many shots she took in each game. In six games, how many shots did she make?

20. If each shot made was worth two points, how many points did she make in all six games?

Monica's Basketball Games

Game	Shots Made	Shots Taken
1	1	3
2	3	4
3	4	7
4	5	11
5	4	7
6	3	8

3.8 Multiplying Two-Digit Numbers

Objective: to multiply two-digit whole numbers by one-digit whole numbers

Five students were in an archery tournament.

Each student shot 32 arrows. How many arrows did the students shoot?

Multiply 32 by 5 to find the answer.

Step 1	Step 2
Multiply the ones. Rename? yes	Multiply the tens.
$\begin{array}{r} 1 \\ 3\overset{}{2} \\ \times\ 5 \\ \hline 0 \end{array}$	$\begin{array}{r} 1 \\ 3\,2 \\ \times\ 5 \\ \hline 160 \end{array}$ $\quad$ $\begin{array}{r} 30 \times 5 =\ \ 150 \\ 2 \times 5 = +\ \ 10 \\ \hline 160 \end{array}$
5×2 ones $= 10$ ones Regroup 10 ones as 1 ten 0 ones.	5×3 tens $= 15$ tens 15 tens $+ 1$ ten $= 16$ tens 16 tens $= 1$ hundred 6 tens

THINK
An estimate is 5×30, or 150.

They shot 160 arrows.

Compare the actual product to the estimate.
Does the answer seem reasonable?

More Examples

A. $\begin{array}{r} 1 \\ 36 \\ \times\ 2 \\ \hline 72 \end{array}$
$\qquad$ **B.** $\begin{array}{r} 1 \\ \$64 \\ \times\ 3 \\ \hline \$192 \end{array}$
$\qquad$ **C.** $\begin{array}{r} 2 \\ 45 \\ \times\ 5 \\ \hline 225 \end{array}$
$\qquad$ **D.** $\begin{array}{r} 4 \\ 65 \\ \times\ 8 \\ \hline 520 \end{array}$

TRY THESE

Estimate. Then multiply.

1. $\begin{array}{r} 11 \\ \times\ 4 \\ \hline \end{array}$
$\qquad$ **2.** $\begin{array}{r} 39 \\ \times\ 2 \\ \hline \end{array}$
$\qquad$ **3.** $\begin{array}{r} \$14 \\ \times\ 7 \\ \hline \end{array}$
$\qquad$ **4.** $\begin{array}{r} 51 \\ \times\ 7 \\ \hline \end{array}$
$\qquad$ **5.** $\begin{array}{r} 27 \\ \times\ 6 \\ \hline \end{array}$
$\qquad$ **6.** $\begin{array}{r} \$45 \\ \times\ 8 \\ \hline \end{array}$

Exercises

Estimate. Then multiply.

1. $\begin{array}{r} 42 \\ \times\ 4 \\ \hline \end{array}$
2. $\begin{array}{r} 49 \\ \times\ 4 \\ \hline \end{array}$
3. $\begin{array}{r} 54 \\ \times\ 7 \\ \hline \end{array}$
4. $\begin{array}{r} 92 \\ \times\ 2 \\ \hline \end{array}$
5. $\begin{array}{r} 50 \\ \times\ 6 \\ \hline \end{array}$

6. $\begin{array}{r} 74 \\ \times\ 6 \\ \hline \end{array}$
7. $\begin{array}{r} 60 \\ \times\ 7 \\ \hline \end{array}$
8. $\begin{array}{r} 53 \\ \times\ 3 \\ \hline \end{array}$
9. $\begin{array}{r} 37 \\ \times\ 6 \\ \hline \end{array}$
10. $\begin{array}{r} 65 \\ \times\ 5 \\ \hline \end{array}$

11. 7×48
12. 4×72
13. 8×78
14. $\$8 \times 57$
15. $\$45 \times 8$
16. $\$24 \times 5$

PROBLEM SOLVING

17. How many hours are in 5 days?

18. How many hours are in 8 days?

19. It takes Gretchen 2 minutes and 12 seconds to load and aim her bow and arrow. How many *seconds* does it take her?

20. If Andrew shoots 23 arrows an hour, how many arrows would he shoot in 6 hours?

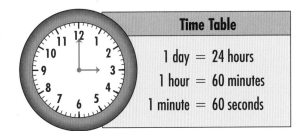

Time Table
1 day = 24 hours
1 hour = 60 minutes
1 minute = 60 seconds

21. Chad practices for 24 minutes on Tuesday, 46 minutes on Wednesday, and 33 minutes on Saturday. How many minutes does he practice in all?

★ 22. Jacob practices archery for 3 hours and 14 minutes. How many *minutes* is this?

CONSTRUCTED RESPONSE

23. The Cougars soccer team won its game. The coach wants to treat the team to ice cream. Each ice cream cone costs 59¢. There are nine boys on the team. Will $5 cover the cost of the ice cream? Explain why or why not.

3.9 Multiplying Three-Digit Numbers

Objective: to multiply three-digit whole numbers by one-digit whole numbers

A swimming pool is 183 feet long. Craig swims 4 lengths of the pool. How far does he swim?

THINK An estimate is 4×200, or 800.

To find the answer, multiply 183 by 4.

Step 1	Step 2	Step 3
Multiply the ones.	Multiply the tens.	Multiply the hundreds.
$18\overset{1}{3}$ $\times \quad 4$ $\overline{\qquad 2}$	$\overset{3}{1}\overset{1}{8}3$ $\times \quad 4$ $\overline{\quad 32}$	$\overset{3}{1}\overset{1}{8}3$ $\times \quad 4$ $\overline{732}$
$4 \times 3 = 12$ 12 ones = 1 ten 2 ones	$(4 \times 8) + 1 = 33$ 33 tens = 3 hundreds 3 tens	$(4 \times 1) + 3 = 7$ 7 hundreds

Craig swims 732 feet.
Compare the actual product to the estimate. Is the answer reasonable?

More Examples

A.
$$\begin{array}{r} 2\overset{1}{3}2 \\ \times \quad 2 \\ \hline 464 \end{array}$$

B.
$$\begin{array}{r} 3\overset{1}{2}6 \\ \times \quad 3 \\ \hline 978 \end{array}$$

C.
$$\begin{array}{r} \overset{2}{1}70 \\ \times \quad 4 \\ \hline 680 \end{array}$$

Do not forget to multiply the zeros.

D.
$$\begin{array}{r} 2\overset{1}{0}3 \\ \times \quad 4 \\ \hline 812 \end{array}$$

TRY THESE

Estimate. Then multiply.

1.
$$\begin{array}{r} 203 \\ \times \quad 9 \\ \hline \end{array}$$

2.
$$\begin{array}{r} 221 \\ \times \quad 4 \\ \hline \end{array}$$

3.
$$\begin{array}{r} \$432 \\ \times \quad 2 \\ \hline \end{array}$$

4.
$$\begin{array}{r} 306 \\ \times \quad 7 \\ \hline \end{array}$$

5.
$$\begin{array}{r} 105 \\ \times \quad 8 \\ \hline \end{array}$$

6.
$$\begin{array}{r} 364 \\ \times \quad 5 \\ \hline \end{array}$$

7.
$$\begin{array}{r} \$120 \\ \times \quad 5 \\ \hline \end{array}$$

8.
$$\begin{array}{r} 336 \\ \times \quad 2 \\ \hline \end{array}$$

9.
$$\begin{array}{r} 147 \\ \times \quad 6 \\ \hline \end{array}$$

10.
$$\begin{array}{r} \$248 \\ \times \quad 3 \\ \hline \end{array}$$

Multiply.

1. 203 × 9	**2.** 340 × 8	**3.** 511 × 7	**4.** 902 × 9	**5.** 810 × 5
6. 999 × 9	**7.** $742 × 7	**8.** $532 × 5	**9.** 946 × 5	**10.** 751 × 2

11. $656 × 3   **12.** $860 × 4   **13.** 489 × 7   **14.** 767 × 9   **15.** 405 × 5

16. 389 × 8   **17.** $368 × 6   **18.** 798 × 7   **19.** 857 × 9

PROBLEM SOLVING

20. There are 365 days in a year. How many days are there in 9 years?

21. Steffi Newcombe earns $632 a week for endorsing a brand of running shoes. How much does she earn in 5 weeks?

22. There are 7 sections of seats in the hockey arena. Each section holds 931 seats. How many seats are in the arena?

★ **23.** Roger bought a season ticket for 9 hockey games. Each ticket would cost $14 if bought one at a time. Roger paid $108 for the season ticket. What was his total savings?

MIND BUILDER

Puzzle

Find the products. Then copy the magic square and write the products in it.

a. 130 × 5   **b.** 7 × 200   **c.** 625 × 2
d. 2 × 850   **e.** 4 × 275   **f.** 125 × 4
g. 475 × 2   **h.** 200 × 4   **i.** 5 × 310

Find the sum of the numbers in each row. Then find the sum of the numbers in each column. All the sums should be the same.

3.10 Multiplying Greater Numbers

Objective: to multiply four-digit (or more) numbers by one-digit whole numbers

There are 8 sections of seats in the sports arena. Each section has 2,105 seats. How many seats are in the arena?

Multiply 2,105 by 8 to find the answer.

THINK
An estimate is 8 × 2,000, or 16,000.

Step 1	Step 2	Step 3	Step 4
Multiply the ones.	Multiply the tens.	Multiply the hundreds.	Multiply the thousands.
$$\begin{array}{r} \overset{4}{2,10\!\!\uparrow\!\!5} \\ \times\quad 8 \\ \hline 0 \end{array}$$	$$\begin{array}{r} \overset{4}{2,1\!\!\uparrow\!\!05} \\ \times\quad 8 \\ \hline 40 \end{array}$$	$$\begin{array}{r} \overset{4}{2,105} \\ \times\quad 8 \\ \hline 840 \end{array}$$	$$\begin{array}{r} \overset{4}{2,105} \\ \times\quad 8 \\ \hline 16,840 \end{array}$$
8 × 5 = 40 40 ones	(8 × 0) + 4 = 4 4 tens	8 × 1 = 8 8 hundreds	8 × 2 = 16 16 thousands

The sports arena has 16,840 seats.

More Examples

A.
$$\begin{array}{r} \overset{1\,1}{3,068} \\ \times\quad 2 \\ \hline 6,136 \end{array} \longrightarrow \begin{array}{r} 3,000 \\ \times\quad 2 \\ \hline 6,000 \end{array}$$

B.
$$\begin{array}{r} \overset{1}{44,251} \\ \times\quad 2 \\ \hline 88,502 \end{array}$$

C.
$$\begin{array}{r} \overset{2}{5,004} \\ \times\quad 5 \\ \hline 25,020 \end{array}$$

TRY THESE

Estimate. Then multiply.

1.
$$\begin{array}{r} 2,513 \\ \times\quad 3 \end{array}$$

2.
$$\begin{array}{r} 4,507 \\ \times\quad 2 \end{array}$$
Why is the estimate greater than the product?

3.
$$\begin{array}{r} 3,162 \\ \times\quad 3 \end{array}$$

4.
$$\begin{array}{r} 7,006 \\ \times\quad 4 \end{array}$$

5. 2,130 × 2	6. 1,237 × 3	7. 12,147 × 4	8. 23,462 × 6	9. 38,799 × 4

Exercises

Multiply.

1. 5,048 × 5	2. 2,222 × 7	3. $7,451 × 9	4. $8,476 × 7	5. 47,040 × 8

6. $4,573 × 7	7. 106,409 × 8	8. $5,947 × 6	9. $3,768 × 9

10. 38,005 × 6

11. $9,305 × 4

Is each answer reasonable? Use your estimation skills to check. Write *yes* or *no*.

12. 2 × 6,298 = 1,256

13. 7,954 × 7 = 55,678

14. 9,846 × 3 = 29,538

15. 3,789 × 9 = 25,101

16. 7 × 9,568 = 66,976

17. 8 × 7,049 = 56,392

PROBLEM SOLVING

18. A kilometer is 1,000 meters. How many meters are in 8 kilometers?

19. Charlie needs to buy 2 sofas. Each sofa costs $278. How much will Charlie spend on the sofas?

20. The concert tickets cost $8 each. If 186 people go to the concert, how much money will the arena make?

21. There are 3,732 books on each floor of the library. The library is four floors. How many books are in the library on all four floors?

TEST PREP

22. David runs 5 miles every day. During the last year, how many miles did he run? (*Remember:* A year has 365 days.)

   a. 3,100 miles   **b.** 2,230 miles   **c.** 1,825 miles   **d.** 1,150 miles

23. MaryLee loves to bake. She bakes 6 dozen muffins each week for her neighbors. How many muffins will she bake during an 8-week time period?

   a. 540 muffins   **b.** 576 muffins   **c.** 500 muffins   **d.** 480 muffins

3.11 Problem-Solving Strategy: Making a Table

Objective: to solve problems by creating a table

Tyler's hobby is baking. He has collected some recipes for making doughnuts. One doughnut recipe calls for 4 cups of flour for every 3 cups of sugar. How many cups of flour are needed for 18 cups of sugar?

1. READ
You need to find the number of cups of flour needed when 18 cups of sugar are used. You know that when 4 cups of flour are used, 3 cups of sugar are used.

2. PLAN
Make a table to organize what you know. Fill in the table. Increase the first row by multiplying by 2, 3, and so on.

3. SOLVE

flour	4	8	12	16	20	24
sugar	3	6	9	12	15	18

24 cups of flour are needed for 18 cups of sugar.

4. CHECK
The solution is correct if 4 and 3 have been multiplied by the same number.

$$4 \times 6 = 24 \checkmark$$

$$3 \times 6 = 18$$

TRY THESE

Copy and complete the table. Add more spaces, if needed. Then solve.

1. In one recipe, you need 4 cups of flour for every egg used. How many eggs are needed if you use 20 cups of flour?

flour	4	8	12	■
eggs	1	2	■	■

2. A lizard has a tail that is twice as long as its body. How long is the body of a lizard with a total length of 21 inches?

body	1	2	3	4	■
tail	2	4	6	■	■
total	3	6	■	■	■

Copy and complete the table. Then solve.

1. Four cans of baked beans cost $1.00. How many cans can you buy for $5.00?

Cost	$1.00	$2.00	$3.00	■	■
Number of cans	4	8	■	■	■

2. Three rails are needed for each section of a fence. How many rails are needed for five sections?

Number of sections	1	2	■	■	■
Number of rails	3	■	■	■	■

Solve. Make a table.

3. San Diego is 1,200 miles away. If Carol drives 400 miles a day, how many days will it take her to make the round-trip?

4. Ben saw squirrels and wild turkeys in the forest. He counted 8 heads and 22 legs. Continue the pattern in the following table. How many turkeys and how many squirrels did he see?

5. A recipe calls for 2 eggs and 3 cups of flour. A chef uses 18 eggs. How many cups of flour are used?

6. In a survey, for every 100 people that liked ice cream, 3 did not. If 9 people disliked ice cream, how many people liked ice cream? How many people were surveyed?

Ben's Forest Finds

Turkeys	Squirrels	Turkey legs	Squirrel legs	Total legs
1	7	2	28	30
2	6	4	24	28
3	5	6	20	26

MIND BUILDER

Product Patterns

Find the products. Look for patterns. When you see a pattern, use it to help you guess the next product.

1. 99×1 2. 99×2 3. 99×3 4. 99×4 5. 99×5

Use your products in problems 2–5 to answer these questions.

6. a. What is the sum of the first and last digits in each product?

   b. What is the middle digit in each product?

7. Multiply 99 by 6, 7, 8, 9, and 10. Try other combinations of factors and 9s.

LANGUAGE and CONCEPTS

Write the letter of the word or number(s) that best completes each sentence.

1. Some multiples of 7 are _____.

2. In $50 \times 70 = 3,500$, 3,500 is called the _____.

3. The _____ Property of Multiplication is shown by $46 \times 7 = 7 \times 46$.

4. If 0 is a factor, the product is _____.

5. Some multiples of 6 are _____.

6. In $41 \times 3 = 123$, 41 and 3 are called _____.

7. The _____ Property is shown here: $14 \times 3 = (10 \times 3) + (4 \times 3)$.

a. product
b. 0
c. 6, 12, 18, and 24
d. Commutative
e. 0, 7, 14, and 21
f. factors
g. 0, 5, 10, and 15
h. Associative
i. Distributive

SKILLS and PROBLEM SOLVING

Write the multiplication facts shown by each picture. (Section 3.1)

8.

9.

Name the property shown here. (Sections 3.2–3.3)

10. $15 \times 12 = 12 \times 15$

11. $31 \times 5 = (30 \times 5) + (1 \times 5)$

12. $392 \times 0 = 0$

13. $5,291 \times 1 = 5,291$

List the first six multiples of the following numbers. (Section 3.4)

14. 5

15. 8

16. 6

List all the factors for the following numbers. (Section 3.4)

17. 49

18. 36

19. 60

Find the missing factor. (Section 3.5)

20. $8 \times \blacksquare = 4{,}800$ **21.** $\blacksquare \times 300 = 2{,}100$ **22.** $7 \times \blacksquare = 56{,}000$

How many different combinations are possible? Make a list. (Section 3.6)

23. vanilla, chocolate, or strawberry ice cream; hot fudge, berry, or caramel topping

24. milk or water; cheeseburger, hot dog, or chicken fingers; fries or apple slices

Estimate by rounding. (Section 3.7)

25.
$$\begin{array}{r} 21 \\ \times\ 2 \\ \hline \end{array}$$

26.
$$\begin{array}{r} 935 \\ \times\ 5 \\ \hline \end{array}$$

27.
$$\begin{array}{r} 1{,}601 \\ \times\ 7 \\ \hline \end{array}$$

28.
$$\begin{array}{r} 4{,}278 \\ \times\ 5 \\ \hline \end{array}$$

Multiply. (Sections 3.8–3.10)

29.
$$\begin{array}{r} 147 \\ \times\ 8 \\ \hline \end{array}$$

30.
$$\begin{array}{r} 121 \\ \times\ 7 \\ \hline \end{array}$$

31.
$$\begin{array}{r} \$313 \\ \times\ 4 \\ \hline \end{array}$$

32.
$$\begin{array}{r} 497 \\ \times\ 9 \\ \hline \end{array}$$

33.
$$\begin{array}{r} \$235 \\ \times\ 7 \\ \hline \end{array}$$

34.
$$\begin{array}{r} 6{,}754 \\ \times\ 6 \\ \hline \end{array}$$

35. $\$85 \times 6$ **36.** 23×4 **37.** 46×7 **38.** 36×5

39. $2{,}502 \times 7$ **40.** $3{,}456 \times 5$ **41.** $5 \times \$285$ **42.** 4×96

43. $6 \times \$4{,}872$ **44.** $4 \times 1{,}924$ **45.** 470×9 **46.** $3{,}756 \times 7$

Solve. Make a table. (Section 3.11)

47. The community neighborhood wanted to put in some swing sets. Each swing set costs $133. How much would 5 swing sets cost?

48. On a road trip, Mark traveled 672 miles per day. Make a table to show how far he traveled during his 8-day trip.

CHAPTER 3 TEST

Multiply.

1.	8 × 4	**2.**	5 × 2	**3.**	7 × 6	**4.**	0 × 19	**5.**	8 × 90
6.	12 × 6	**7.**	20 × 6	**8.**	15 × 8	**9.**	210 × 1	**10.**	400 × 9
11.	120 × 4	**12.**	1,050 × 5	**13.**	621 × 7	**14.**	5,370 × 4	**15.**	1,967 × 2

Copy and complete.

16. $34 \times 5 = 5 \times \blacksquare$

17. $37 \times 9 = (\blacksquare \times 9) + (\blacksquare \times 9)$

18. $(70 \times 8) + (9 \times 8) = \blacksquare \times 8$

Write the first five multiples for each number.

19. 8   **20.** 5   **21.** 9   **22.** 7

Find the missing factor.

23. $3 \times \blacksquare = 120$ **24.** $50 \times \blacksquare = 450$ **25.** $9 \times \blacksquare = 900$

Solve.

26. There are 5 players on a basketball team. How many players are on 7 teams?

27. The tennis club was preparing for its summer tennis camps. The coach put 7 tennis balls in each basket. How many tennis balls are in 66 baskets?

28. For every 5 tickets you buy for the movies, you receive 2 free tickets. How many free tickets do you receive if you buy 15 tickets?

29. Tyler has soccer practice twice a week with a game on each Saturday. Make a table to show how many times he plays soccer during his 8-week soccer season.

30. In basketball, an assist is made when one player passes the ball to another player who scores a goal. Betsy, Ariana, Grace, Yolanda, and Lucia are on a team. Make a list of all the possible assists. (Betsy passing to Ariana is one possible assist.)

The Sieve of Eratosthenes

Eratosthenes was a Greek mathematician, geographer, and astronomer who lived from 276 B.C.–194 B.C. He is noted for many accomplishments that were remarkable for that time period. Among his scientific feats are the following:

- He created a system of latitude and longitude.
- He made a calendar with leap years.
- He was the first known person to calculate the circumference of the Earth—one that was remarkably close to the actual measurement.
- He designed a star catalogue.
- He created a way of finding **prime numbers** called the "Sieve of Eratosthenes."

A **sieve** is a tool used to allow items, such as flour or water, to sift or funnel through. Go on your own historical exploration, sifting through numbers in search of prime numbers using the Sieve of Eratosthenes. To do so, use a number grid and the following directions. Can you follow in Eratosthenes' footsteps to find the prime numbers?

1. Copy the following number grid.

2. Use a blue crayon or colored pencil to circle the number 2. Then use blue to color all the squares of numbers that are multiples of 2. The number 2 is your first prime number. Keep a list of the prime numbers you find.

3. Next, use a red crayon to circle the next number that is not colored. Here it is the number 3. This is your next prime number. Add it to your list. Then color all of the multiples of 3 red. (Some squares will already have a color in it; it is up to you if you color it again.)

4. Use a green crayon to circle the next number that is not colored. Add that number to your list, and then color the spaces for all the numbers that are multiples of this number.

5. Continue this process of circling the next uncolored number and coloring the multiples. All of the circled numbers are your prime numbers.

1	2	3	4	5	6	7	8	9	10
11	12	13	14	15	16	17	18	19	20
21	22	23	24	25	26	27	28	29	30
31	32	33	34	35	36	37	38	39	40
41	42	43	44	45	46	47	48	49	50
51	52	53	54	55	56	57	58	59	60
61	62	63	64	65	66	67	68	69	70
71	72	73	74	75	76	77	78	79	80
81	82	83	84	85	86	87	88	89	90
91	92	93	94	95	96	97	98	99	100
101	102	103	104	105	106	107	108	109	110

CUMULATIVE TEST

1. Which number is less than 3,670?
   a. 3,760
   b. 3,402
   c. 3,980
   d. none of the above

2. Estimate by rounding.
   $$643 - 326$$
   a. 400
   b. 800
   c. 900
   d. none of the above

3. $7 \times 8 =$ _____
   a. 56
   b. 64
   c. 65
   d. none of the above

4.
   $$4,379$$
   $$6,582$$
   $$+ \ 7,894$$
   a. 12,452
   b. 15,762
   c. 17,645
   d. 18,855

5. $72,093 - 5,496 =$ _____
   a. 66,597
   b. 66,617
   c. 73,403
   d. none of the above

6. Which number is a multiple of 6?
   a. 28
   b. 30
   c. 64
   d. none of the above

7. John has delivered 53 papers. He has 39 more papers to deliver. How many papers will he deliver in all?
   a. 14 papers
   b. 82 papers
   c. 92 papers
   d. 39 papers

8. Simon's Pet Store has 34 angelfish, 13 dogs, 12 cats, 19 goldfish, 9 hamsters, 5 hermit crabs, 4 rabbits, and 3 catfish. How many fish does Simon's Pet Store have in stock?
   a. 39 fish   b. 56 fish
   c. 75 fish   d. 99 fish

9. Which number is a factor of 54?
   a. 5
   b. 4
   c. 9
   d. none of the above

10. Carlos types 2 pages in 15 minutes. How many pages can he type in 90 minutes?

Pages	2	4	6	■
Minutes	15	30	■	■

   a. 8 pages
   b. 10 pages
   c. 12 pages
   d. 14 pages

CHAPTER 4

Multiplying by Two-Digit Numbers

Johanna Bazin
Texas

4.1 Mental Math: Multiples of 10

Objective: to multiply whole numbers by multiples of 10, 100, and 1,000; to use multiples of 10 to help estimate

Some city buildings are made of bricks. Imagine a wall of 30 rows of bricks with 60 bricks in each row. To find the total number of bricks, multiply 30 and 60.

tens × tens = hundreds

3 tens × 6 tens

Count the hundreds.
18 hundreds

So, 30 × 60 = 1,800.

There are 1,800 bricks in all.

Look for patterns and basic facts.

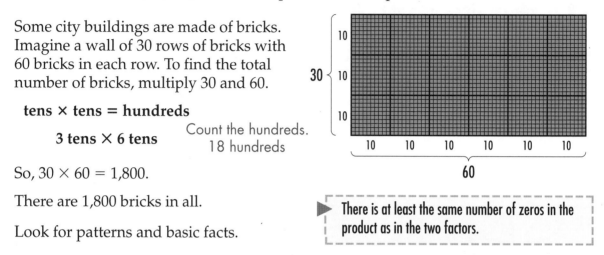

▶ There is at least the same number of zeros in the product as in the two factors.

Examples

- Multiples of 10 can help you estimate products.
- Round each factor to its greatest place-value position.
- Estimate if you see the clue word *about*.

A.
$$\begin{array}{r} 400 \\ \times\ \ 90 \\ \hline 36,000 \end{array}$$
THINK
$$\begin{array}{r} 4 \\ \times\ 9 \\ \hline 36 \end{array}$$

B. 3,000 × 60 = 180,000
   3 × 6 = 18

What pattern do you notice with the zeros?

C. 50 × 800 = 40,000
   5 × 8 = 40

Does this product follow the pattern? Why or why not?

More Examples

D.
$$\begin{array}{r} 49 \\ \times\ 24 \\ \hline \end{array} \rightarrow \begin{array}{r} 50 \\ \times\ 20 \\ \hline 1,000 \end{array}$$

E.
$$\begin{array}{r} 3,429 \\ \times\ \ \ 36 \\ \hline \end{array} \rightarrow \begin{array}{r} 3,000 \\ \times\ \ \ \ 40 \\ \hline 120,000 \end{array}$$

F.
$$\begin{array}{r} 379 \\ \times\ 89 \\ \hline \end{array} \rightarrow \begin{array}{r} 400 \\ \times\ \ 90 \\ \hline 36,000 \end{array}$$

TRY THESE

Multiply mentally.

1. 30 × 40   **2.** 30 × 400   **3.** 30 × 4,000

Estimate only.

4.   83   80   **5.**   45   50   **6.**   562   **7.**   7,574
       × 21 × 20   × 84 × 80   × 38   × 61

Exercises

Multiply mentally.

1. 20 × 900   **2.** 70 × 4,000   **3.** 9,000 × 600   **4.** 600 × 60,000

Estimate only.

5.   184   **6.**   325   **7.**   2,777   **8.**   6,385   **9.**   8,429
       × 92   × 41   ×   44   ×   83   ×   33

10. 25 × 945   **11.** 40 × 2,808   **12.** 82 × 5,538   **13.** 17 × 36,372

PROBLEM SOLVING

Use mental math to solve.

14. Mr. Rosenberg can drive his car 40 miles on 1 gallon of gasoline. His car has a 20-gallon tank. How far can he drive on a full tank of gas?

15. A bubble gum machine can hold 3,432 pieces of bubble gum. If the machine is filled 36 times a year, about how many pieces of bubble gum are chewed in a year?

MIND BUILDER

What Comes Next?

Find the next term in each sequence.

1. 4, 9, 14, 19, 24, 29, ___   **2.** 1, 4, 9, 16, 25, 36, ___

3. 1, 1, 2, 3, 5, 8, 13, ___   **4.** 1, 2, 6, 24, 120, ___

4.2 Multiplying Two-Digit Numbers

Objective: to multiply two-digit whole numbers without regrouping

Tickets for the symphony cost $21 for one person. To find the cost for 13 tickets, multiply $21 by 13.

THINK An estimate is 10×20, or 200.

Step 1	Step 2	Step 3
Multiply by the ones.	Multiply by ten.	Add.
$\begin{array}{r} \$21 \\ \times\ 13 \\ \hline 63 \end{array}$ $3 \times 21 = 63$ The cost for 3 tickets is $63.	$\begin{array}{r} \$21 \\ \times\ 13 \\ \hline 63 \\ 210 \\ \hline \end{array}$ 1 ten × 21 = 21 tens The cost for 10 tickets is $210.	$\begin{array}{r} \$21 \\ \times\ 13 \\ \hline 63 \\ +\ 210 \\ \hline \$273 \end{array}$ Add to find the total cost.

In Step 2, the second partial product has a multiple of ten as one of its factors, which means it must end in zero. The cost for 13 tickets is $273.

Is the answer reasonable? Compare it to the estimate.

More Examples

A.
$\begin{array}{r} 22 \\ \times\ 34 \\ \hline 88 \\ +\ 660 \\ \hline 748 \end{array}$ 4×22 3 tens × 22

B.
$\begin{array}{r} 31 \\ \times\ 11 \\ \hline 31 \\ +\ 310 \\ \hline 341 \end{array}$ 1×31 1 ten × 31

C.
$\begin{array}{r} 30 \\ \times\ 22 \\ \hline 60 \\ +\ 600 \\ \hline 660 \end{array}$ 2×30 2 tens × 30

TRY THESE

Multiply.

1. $\begin{array}{r} 83 \\ \times\ 11 \end{array}$ $\begin{array}{l} 1 \times 83 \\ 1 \text{ ten} \times 83 \end{array}$

2. $\begin{array}{r} 41 \\ \times\ 18 \end{array}$ $\begin{array}{l} 8 \times 41 \\ 1 \text{ ten} \times 41 \end{array}$

3. $\begin{array}{r} 21 \\ \times\ 32 \end{array}$

4. $\begin{array}{r} 40 \\ \times\ 21 \end{array}$

Exercises

Multiply.

1. $\begin{array}{r} 33 \\ \times\ 23 \end{array}$

2. $\begin{array}{r} 90 \\ \times\ 16 \end{array}$

3. $\begin{array}{r} 42 \\ \times\ 22 \end{array}$

4. $\begin{array}{r} 30 \\ \times\ 26 \end{array}$

PROBLEM SOLVING

5. While staying at a hotel, you attend a concert with friends. Find the price of 11 concert tickets.

$12 CONCERT Jan 24

6. Do 3 cans of tennis balls cost more or less than $6? How much more or less?

SALE $1.98

CONSTRUCTED RESPONSE

7. Estimate 24 × 32.

   Is 832 a reasonable estimate? Explain your reasoning.

8. Create a number story illustrating the equation 52 × 36 = 1,872.

   Then explain the steps you take to solve this problem.

TEST PREP

9. The answer to a multiplication problem is called the _____.

   a. sum   **b.** factor   **c.** dividend   **d.** product

4.3 Multiplying Two-Digit Numbers with Regrouping

Objective: to multiply two-digit whole numbers with regrouping

An office building is 46 stories high. Each story has 32 windows. To find the total number of windows, multiply 32 by 46.

THINK
An estimate is
50 × 30, or 1,500.

Step 1	Step 2	Step 3
Multiply by the ones. Regroup.	Multiply by the tens.	Add.
$$\begin{array}{r} \overset{1}{}32 \\ \times\ 46 \\ \hline 192 \end{array}$$ 6 × 32	$$\begin{array}{r} \overset{1}{3}2 \\ \times\ 46 \\ \hline 192 \\ +1{,}280 \end{array}$$ 4 tens × 32	$$\begin{array}{r} \overset{1}{3}2 \\ \times\ 46 \\ \hline 192 \\ +1{,}280 \\ \hline 1{,}472 \end{array}$$

The building has 1,472 windows.
Is the answer reasonable?

▶ Be careful when you need to regroup your ones or tens.

More Examples

Estimate, then find the product.

A.
$$\begin{array}{r} 32 \\ \times\ 48 \\ \hline \end{array} \quad \begin{array}{r} 30 \\ \times\ 50 \\ \hline 1{,}500 \end{array} \quad \begin{array}{r} \overset{1}{3}2 \\ \times\ 48 \\ \hline 256 \\ +1{,}280 \\ \hline 1{,}536 \end{array}$$

B.
$$\begin{array}{r} 43 \\ \times\ 38 \\ \hline \end{array} \quad \begin{array}{r} 40 \\ \times\ 40 \\ \hline 1{,}600 \end{array} \quad \begin{array}{r} \overset{2}{4}3 \\ \times\ 38 \\ \hline 344 \\ +1{,}290 \\ \hline 1{,}634 \end{array}$$ 8 × 43

TRY THESE

Multiply.

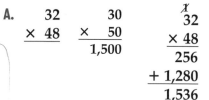

1. $\begin{array}{r} 44 \\ \times\ 25 \\ \hline \end{array}$ 5 × 44 2 tens × 44

2. $\begin{array}{r} 57 \\ \times\ 14 \\ \hline \end{array}$

3. $\begin{array}{r} 24 \\ \times\ 27 \\ \hline \end{array}$

4. $\begin{array}{r} 52 \\ \times\ 38 \\ \hline \end{array}$

Exercises

Do you need to regroup? Write *yes* or *no*. Do not solve.

1. 89×17 **2.** 56×14 **3.** 60×28 **4.** 52×46

Multiply.

5. $\begin{array}{r} 44 \\ \times\ 28 \\ \hline \end{array}$
6. $\begin{array}{r} 79 \\ \times\ 18 \\ \hline \end{array}$
7. $\begin{array}{r} 62 \\ \times\ 17 \\ \hline \end{array}$
8. $\begin{array}{r} 98 \\ \times\ 12 \\ \hline \end{array}$
9. $\begin{array}{r} 24 \\ \times\ 26 \\ \hline \end{array}$
10. $\begin{array}{r} 92 \\ \times\ 28 \\ \hline \end{array}$

11. $\begin{array}{r} 31 \\ \times\ 77 \\ \hline \end{array}$
12. $\begin{array}{r} 53 \\ \times\ 35 \\ \hline \end{array}$
13. $\begin{array}{r} 87 \\ \times\ 13 \\ \hline \end{array}$
14. $\begin{array}{r} 52 \\ \times\ 46 \\ \hline \end{array}$
15. $\begin{array}{r} 82 \\ \times\ 47 \\ \hline \end{array}$
16. $\begin{array}{r} 74 \\ \times\ 24 \\ \hline \end{array}$

Solve.

★**17.** Add is to sum as multiply is to _____.

★**18.** What is the product when you multiply 21 by the sum of 23 and 45?

PROBLEM SOLVING

19. A camera store orders 41 cameras. Each camera costs $62. What is the total cost?

20. A school has 32 buses. Each bus seats 46 students. How many students can be seated in all?

21. A school has 24 desks in each room. There are 13 rooms in the school. How many desks are in the school?

22. Maria bought 15 apples, 12 pears, and 20 oranges. She gave 9 oranges away. How many pieces of fruit does she have now?

MID-CHAPTER REVIEW

Multiply mentally.

1. 10×70 **2.** 40×600 **3.** 10×78 **4.** 70×70

Estimate.

5. 58×62 **6.** 41×503 **7.** 94×389 **8.** $6,409 \times 97$

Multiply.

9. 14×62 **10.** 75×18 **11.** 30×86 **12.** $14 \times \$95$

4.4 Problem-Solving Strategy: Guessing and Checking

Objective: to solve problems using the guess-and-check method

Jan bought some roses and some daisies. She told Todd that she bought 9 flowers and spent $13. Here is how Todd figured out how many of each kind she bought.

Some problems can be solved by making a guess and then checking it. Then make a closer guess and check it.

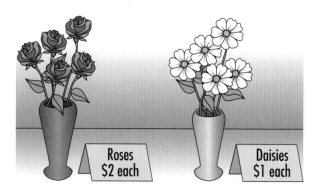

Roses
$2 each

Daisies
$1 each

1. READ You need to find the number of roses and daisies Jan bought.
You know how much she spent, how many flowers she bought, and how much each flower cost.

2. PLAN Use the **guess-and-check method**.
- Make a careful guess.
- Check to see if the guess is correct.
- If the guess is not correct, use the result to improve the next guess.
- Repeat this plan until you find the answer.

3. SOLVE

	Number of Roses	Cost of Roses	Number of Daisies	Cost of Daisies	Total Cost	
1st guess	5	$10	9 − 5 = 4	$4	$14	too much
2nd guess	3	$6	9 − 3 = 6	$6	$12	too little
3rd guess	4	$8	9 − 4 = 5	$5	$13	correct

She bought 4 roses and 5 daisies.

4. CHECK Compare the answer to the original problem.

$$\begin{array}{r} \$2 \\ \times\ 4 \\ \hline \$8 \end{array}$$ cost of 4 roses

$$\begin{array}{r} \$1 \\ \times\ 5 \\ \hline \$5 \end{array}$$ cost of 5 daisies

So, 9 flowers cost $8 + $5, or $13.

TRY THESE

Use the guess-and-check method to solve.

1. When a certain number is multiplied by itself, the product is 225. Find the number.

2. The difference between two numbers is 8. The product is 105. Find the two numbers.

Solve

1. Pencils cost 5¢. Erasers cost 2¢. Amber paid 24¢ for a total of six items. How many of each did she buy?

2. Brice can buy film in rolls of 12 or 20 frames. He buys 5 rolls and gets 76 frames. How many rolls of each film did he buy?

3. Frank earns $1 the first day, $2 the second day, $4 the third day, and so on. How much does he earn the sixth day?

4. Dora buys some clay pots and some ferns. She spent $45. Using the prices shown, how many clay pots and how many ferns did she buy?

5. Crayons are sold in boxes of 12 or 16. Nellie bought 5 boxes and got 76 crayons. How many boxes of 12 crayons and how many boxes of 16 crayons did Nellie buy?

MIXED REVIEW

Tell how many in all.

6. 10 sets of 20

7. 3 sets of 3,000

8. 9 sets of 40

9. 80 sets of 10

10. 56 sets of 100

11. 35 sets of 1 million

Complete.

12. 2,346 + 391

13. 2,346 − 391

14. 391 × 6

15. What do you notice about problems 12 and 13?

Problem Solving

How Many Rounds of Hopscotch?

Terry and Joella drew a hopscotch board in their driveway. As they were playing, they wondered what would happen if they started adding the numbers that they were jumping on, but not adding in the number the stone was on. They wondered what total they would have after three full rounds, with the first round not adding in the 1, the second round not adding in the 2, and the third round not adding in the 3. Terry predicted 200; Joella predicted 300. Were either of the friends right? What number would they reach after three rounds of hopscotch?

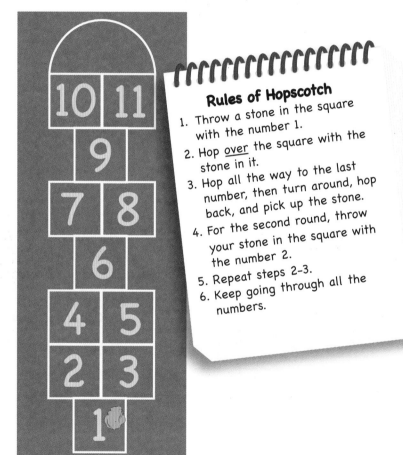

Rules of Hopscotch

1. Throw a stone in the square with the number 1.
2. Hop <u>over</u> the square with the stone in it.
3. Hop all the way to the last number, then turn around, hop back, and pick up the stone.
4. For the second round, throw your stone in the square with the number 2.
5. Repeat steps 2–3.
6. Keep going through all the numbers.

Extension

1. How many rounds would it take to get to 750 points?

2. How many points would they get if they went twelve full rounds?

CUMULATIVE REVIEW

Order each group of numbers from greatest to least.

1. 2,152   2,209   2,125   **2.** 517   5,721   5,210   **3.** 9,642   9,641   9,542

Write the fact family for each group of numbers.

4. 4, 5, 9   **5.** 2, 6, 8   **6.** 7, 12, 5

Estimate.

7.	**8.**	**9.**	**10.**	**11.**
8,124 − 4,677	8,056 + 1,495	33 × 7	671 × 8	2,632 × 5

Compute.

12.	**13.**	**14.**	**15.**	**16.**
3,641 + 8,093	5,958 − 3,967	263 + 532	$37.42 + 15.98	6,834 − 6,703

17. 5 × $9.27   **18.** 13 × 20   **19.** 36 × 40

20. 13 × 46 =   **21.** 24 × 83   **22.** 20 × 30 × 70

Write each number in words and in expanded form.

23. 7,896

24. 45,000,001

25. 56,874

26. 9,999,097

Solve. Use the table.

27. During which month was the rainfall the greatest?

28. Which months had more than a half of a foot of rain? *Remember:* There are 12 inches in 1 foot.

29. Find the total rainfall for May, June, and July.

Monthly Rainfall in Inches	
January	2
February	3
March	5
April	4
May	7
June	6
July	5
August	5
September	9
October	7
November	6
December	4

4.5 Multiplying Greater Numbers

Objective: to multiply multidigit whole numbers by two-digit numbers both with and without regrouping

A delivery truck travels the same route of 124 miles each day. How many miles does it travel in 12 days?

Multiply 124 by 12.

THINK
An estimate is 100×10, or 1,000.

Step 1	Step 2	Step 3
Multiply by the ones.	Multiply by the tens.	Add.
$\begin{array}{r} 124 \\ \times\ \ 12 \\ \hline 248 \end{array}$ 2×124	$\begin{array}{r} 124 \\ \times\ \ 12 \\ \hline 248 \\ 1{,}240 \end{array}$ 1 ten $\times$ 124	$\begin{array}{r} 124 \\ \times\ \ 12 \\ \hline 248 \\ +1{,}240 \\ \hline 1{,}488 \end{array}$

The truck travels 1,488 miles.   How do you know if your answer makes sense?

Another Example

You may need to regroup when multiplying by the ones and the tens.

Multiply 296 by 32.

Step 1	Step 2	Step 3
Multiply by the ones.	Multiply by the tens.	Add.
$\begin{array}{r} \overset{x\ x}{296} \\ \times\ \ 32 \\ \hline 592 \end{array}$	$\begin{array}{r} \overset{2\ 1}{\overset{x\ x}{296}} \\ \times\ \ 32 \\ \hline 592 \\ 8{,}880 \end{array}$	$\begin{array}{r} \overset{2\ 1}{\overset{x\ x}{296}} \\ \times\ \ 32 \\ \hline 592 \\ +8{,}880 \\ \hline 9{,}472 \end{array}$
1. $2 \times 6 = 12$ (12 ones) Regroup as 1 ten 2 ones. 2. $(2 \times 9) + 1 = 19$ (19 tens) Regroup as 1 hundred 9 tens. 3. $(2 \times 2) + 1 = 5$ (5 hundreds)	1. $3 \times 6 = 18$ (18 tens) Regroup as 1 hundred 8 tens. 2. $(3 \times 9) + 1 = 28$ (28 hundreds) Regroup as 2 thousands 8 hundreds. 3. $(3 \times 2) + 2 = 8$ (8 thousands)	

The answer is 9,472.

Multiply.

1. 102
    × 18

2. 811
    × 24

3. $430
    × 31

4. 234
    × 27

Multiply.

1. 109
    × 69

2. 444
    × 23

3. $482
    × 34

4. 824
    × 72

5. 564
    × 33

6. 1,782
    × 65

7. 2,045
    × 82

8. 3,674
    × 89

9. $8,754
    × 72

10. 6,489
    × 98

Solve.

★11. Addend is to sum as _____ is to product.

★12. Compare using <, >, or =.
78 × 6,475 ● 4,810 × 84

PROBLEM SOLVING

13. If a case of nails holds 6,832 nails, how many nails would be in 31 cases?

★14. Mrs. Marburg makes $2,367 a month. How much does she make in 2 years?

MIND BUILDER

Hink Pinks

A hink pink is a riddle where the answer is a two-word answer of rhyming words. For example, a <u>swampy canine</u> would be a **bog dog**. Use your multiplication skills and the letter codes to answer this hink pink.

A	B	C	E	F	H	M	O	R	T	Z
252	817	1,102	729	672	1,836	1,200	1,512	308	836	1,600

What you are in when you are swimming in numbers?

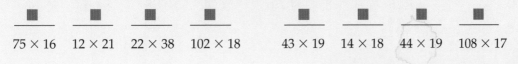

___ ___ ___ ___   ___ ___ ___ ___
75 × 16   12 × 21   22 × 38   102 × 18   43 × 19   14 × 18   44 × 19   108 × 17

4.6 Multiplying with Money

Objective: to multiply money amounts by two-digit whole numbers

A street vendor sold 37 hot dogs for $1.25 each. Multiply $1.25 by 37 to find how much money the street vendor collected.

THINK An estimate is 40 × $1, or $40.

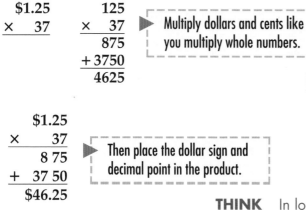

```
  $1.25        125
×    37     ×   37    ▶  Multiply dollars and cents like
               875        you multiply whole numbers.
            + 3750
              4625
```

```
  $1.25
×    37
   8 75     ▶  Then place the dollar sign and
+  37 50       decimal point in the product.
 $46.25
```

The street vendor collected $46.25.

THINK In looking at the answer and the estimate, the answer seems reasonable.

More Examples

```
A.    $3.08
   ×     64   60 × $3
      12 32
   + 184 80
    $197.12
```

```
B.    $0.67
   ×     43
       2 01
   + 26 80
    $28.81
```

```
C.   $28.65
   ×     27
     200 55
   + 573 00
    $773.55
```

TRY THESE

Multiply.

1. $2.50
   ×   16

2. $1.28
   ×   24

3. $0.56
   ×   38

4. $0.98
   ×   20

5. $5.63
   ×   32

Exercises

Multiply.

1. $0.85 × 26

2. $0.46 × 71

3. $0.32 × 95

4. $0.74 × 39

5. $1.20 × 62

6. $1.35 × 25

7. $1.88 × 51

8. $1.99 × 36

9. $1,289 × 18

10. $1,899 × 28

11. $1,709 × 38

12. $1,089 × 48

13. $5,890 × 3

14. $25,090 × 5

15. $35,900 × 7

16. $55,799 × 9

PROBLEM SOLVING

17. Jose earned $1.30 an hour as a delivery boy for the family business. One month he worked 46 hours. How much did he earn that month?

★ **18.** Kenly earns $8.90 an hour. If she works 56 hours over a 2-week pay period, how much does she earn?

19. The city florist made 40 corsages for the holiday. If 8 corsages were sold yesterday and 15 today, how many corsages are left?

CONSTRUCTED RESPONSE

20. At Josie's lemonade stand, she sold 23 glasses of lemonade for $0.25 each and 29 cookies for $0.35 each.

a. Which item brought in more money? How much more?

b. What was the total intake?

c. Josie believed that if she charged $0.70 for each cookie, she would have doubled her total sales. Her business manager disagreed, saying that the lemonade price would have also needed to be raised in order for this to happen. Explain how the business manager was correct.

MIXED REVIEW

Find the sum or difference.

21. 56,789
       + 7,193

22.   8,000
       − 783

23. $25.58
       − 8.77

24. $19.95
       + 1.20

4.7 Problem-Solving Application: Shopping Cart

Objective: to solve multiplication problems involving money and data sources

Annie goes to the produce stand. She needs 18 apples to make applesauce. (The prices show the cost per item.) How much money will she need?

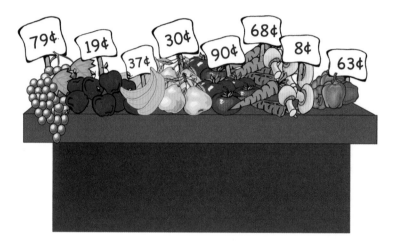

| 1. READ | Find the price of 18 apples. Look at the picture to find the price of one apple. |

| 2. PLAN | Since each apple is the same price, multiply 19¢ by 18. |

Estimate. **20 × 20 = $4.00**

| 3. SOLVE |

$$
\begin{array}{r}
19¢ \\
\times\ 18 \\
\hline
152 \\
+\ 190 \\
\hline
342¢ \text{ or } \$3.42
\end{array}
$$

The price of 18 apples is $3.42.

| 4. CHECK | Compare the answer to the estimate. Since the estimate is $4.00, the answer is reasonable. |

TRY THESE

Complete the table.

Produce	How Many?	Price for Each	Total Price
Green peppers	23	63¢	1.
Carrots	12	68¢	2.
Mushrooms	16	8¢	3.
Total Amount of Order			4.

Solve

Complete the table. Use the produce information from the previous page.

Produce	How Many?	Price for Each	Total Price
Apples	12	19¢	1.
Grapes	3	2.	3.
Tomatoes	14	4.	5.
Pears	7	6.	7.
Bananas	5	8.	9.
Total Amount of Order			10.

Solve using the produce information from the previous page.

11. Fran needs to buy 4 of each vegetable to make a stew. How much money will she spend?

12. The grocery store sells carrots for 75¢ each. How much do you save by buying 9 carrots from the produce stand?

CHAPTER 4 REVIEW

LANGUAGE and CONCEPTS

Write the letter of the correct word that best completes each sentence.

1. One way to estimate is to first round each factor to its _____ place-value position. Then multiply.

2. 40 is the same as 4 _____.

3. 500 is the same as 5 _____.

4. In $47 \times 2,897 = 136,159$, the number 47 is called a _____.

5. Multiply is to _____ as subtract is to difference.

a. greatest
b. hundreds
c. factor
d. product
e. tens

SKILLS and PROBLEM SOLVING

Multiply mentally. Use paper and pencil to write only the products. (Section 4.1)

6. 20×40

7. 30×50

8. $40 \times 9,000$

9. 60×700

10. $90 \times 30,000$

11. 800×500

12. $2,000 \times 600$

13. $6,000 \times 5,000$

Estimate. (Section 4.1)

14. 29
   × 65

15. 21
   × 12

16. 935
   × 58

17. 419
   × 42

18. 390
   × 62

19. 429
   × 34

20. 1,600
   × 78

21. 4,278
   × 55

Multiply. (Sections 4.2–4.3 and 4.5–4.6)

22. 45
    × 13

23. 238
    × 19

24. 48
    × 40

25. 82
    × 90

26. $25
    × 30

27. 396
    × 50

28. $1.57
    × 62

29. 289
    × 97

30. 1,694
    × 73

31. $27.15
    × 17

32. $47.59
    × 76

33. 6,479
    × 89

34. $377.59
    × 97

35. 5,749 × 37

36. 50 × $37.45

37. 46 × 38,327

Solve using the guess-and-check method and the menu provided. (Sections 4.4 and 4.7)

38. Bryan ordered a sandwich for himself and one for his brother. He spent $6.98. What two sandwiches did he buy?

39. Ken bought six of the same sandwiches for his football buddies. He spent $29.34. What sandwiches are he and his pals going to eat during the game?

★40. Ms. Sweeley bought lunch for her fellow workers. She bought three different kinds of sandwiches: 5 of one kind, 3 of another kind, and 2 of a third kind. She spent $58.50. What sandwiches did she order?

Menu

SANDWICHES

Roast Beef	$4.89
Turkey	$3.99
Chicken Salad	$4.79
Shrimp Salad	$7.99
Tuna Salad	$2.99

CHAPTER 4 TEST

Estimate.

1. $\begin{array}{r} 93 \\ \times\ 42 \\ \hline \end{array}$
2. $\begin{array}{r} 46 \\ \times\ 75 \\ \hline \end{array}$
3. $\begin{array}{r} 218 \\ \times\ 21 \\ \hline \end{array}$
4. $\begin{array}{r} 785 \\ \times\ 37 \\ \hline \end{array}$

Multiply.

5. $\begin{array}{r} 14 \\ \times\ 10 \\ \hline \end{array}$
6. $\begin{array}{r} 27 \\ \times\ 30 \\ \hline \end{array}$
7. $\begin{array}{r} 781 \\ \times\ 20 \\ \hline \end{array}$
8. $\begin{array}{r} 32 \\ \times\ 24 \\ \hline \end{array}$

9. $\begin{array}{r} 986 \\ \times\ 65 \\ \hline \end{array}$
10. $\begin{array}{r} 437 \\ \times\ 93 \\ \hline \end{array}$
11. $\begin{array}{r} 2,184 \\ \times\ 56 \\ \hline \end{array}$
12. $\begin{array}{r} 4,306 \\ \times\ 74 \\ \hline \end{array}$

13. 0.48×17
14. 9.24×97
15. $46 \times \$13.57$

16. 89.99×25
17. $7,841 \times 68$
18. 40.08×37

Solve.

19. Special windows in an office building downtown cost $2,879 each. What is the cost of ordering 30 such windows?

20. A parking lot has 18 rows of parking spaces. There are 28 spaces in each row. How many spaces are in the lot?

Use the school supplies price list to solve problems 21–24.

21. Lee needs to purchase 8 notebooks. How much will he spend?

22. Sam needs to buy 17 glue sticks for a school project. Will $5.00 be enough money to buy the glue sticks?

23. Mrs. Green bought 15 of the same item. She spent $17.85. What did she buy?

24. Darryl bought 3 different items. He spent 3.08. What 3 items did he buy?

School Supplies	
Crayons	$1.19
Glue sticks	$0.29
Notebooks	$1.57
Ruler	$1.05
Eraser	$0.46

Lattice Multiplication

Lattice multiplication is another way to find products.

Multiply 54 by 62.

Make a diagram. Write the factors as shown.

Use multiplication facts to fill in the boxes.

Add the numbers on each diagonal.

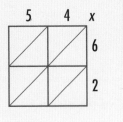 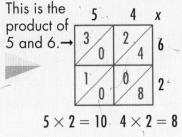 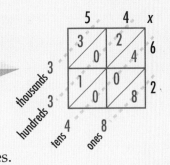

This is the product of 5 and 6.→

$5 \times 2 = 10$ $4 \times 2 = 8$

Read the product along the left and bottom sides.

The product is 3,348.

Multiply 245 by 37.

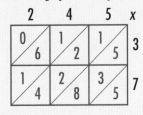 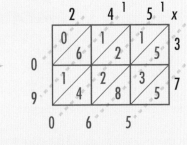

If one sum on a diagonal is more than 9, regroup.

The product is 9,065.

Find each product. Use lattice multiplication.

1.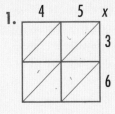

2.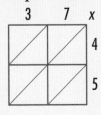

3.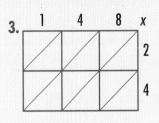

4. 54×23 **5.** 203×31 **6.** 426×73 **7.** 297×49

1. 75,828 + 64,614 = _____
a. 139,442
b. 140,442
c. 161,151
d. none of the above

2. 17 − (34 − 26) = _____
a. 9
b. 10
c. 11
d. none of the above

3. 6,542 − 2,891 = _____
a. 3,651
b. 3,751
c. 4,351
d. none of the above

4. $1.64 × 55 = _____
a. $16.40
b. $86.30
c. $90.20
d. none of the above

5. Choose the best estimate for 28 × 71.
a. 592
b. 1,400
c. 2,100
d. none of the above

6. Choose the number that is the solution to 3 × (23 + 34 + 14).
a. 213
b. 439
c. 788
d. none of the above

7. There are 24 hours in 1 day. How many hours are there in 1 week?
a. 125
b. 137
c. 168
d. none of the above

8. With $30.00, Tami bought a dress, a pair of jeans, and a pair of socks. What was her change?

dress	$14.59
jeans	$ 7.99
socks	$ 0.99

a. $6.43
b. $7.53
c. $23.57
d. none of the above

9. Paul wants to buy some candy from the store. Each piece of candy costs 50¢. Paul has $2.50. How many pieces will he be able to buy if he does not need to worry about tax?
a. 4
b. 6
c. 5
d. none of the above

10. The thousands and the hundreds digits are the same. The ones digit is 2 less than the hundreds digit. The tens digit is the sum of the thousands digit and the hundreds digit. The ones digit is 0. What is the number?
a. 3,340
b. 1,120
c. 2,240
d. none of the above

CHAPTER 5

Statistics and Data

Keech Turner
Calvert Day School

5.1 Collecting, Organizing, and Displaying Data

Objective: to organize collected data using tally charts and line plots

Mr. Sutter's fourth grade students are studying animals. Mr. Sutter suggested that the students collect **data**, or factual information, about their favorite pets. The students' results are found in the table.

The table at the right is a **tally chart**. It is an easy way to keep track of things that are being counted. The middle column of the chart is a visual display of the data. The last column displays the total number of the tally marks for each item counted.

Animal	Tally	Total Number Counted
Cat	ⵑ	5
Dog	ⵑ l	6
Rabbit	ll	2
Parakeet	lll	3
Snake	l	1

Tally marks are displayed in groups of five (four marks with a slash) to make totaling them easier.

Data can also be displayed by using a number line and constructing a **line plot**. Observe how a list of test scores is displayed on a line plot:

85, 90, 70, 60, 85, 60, 85, 60, 90, 85, 40, 60

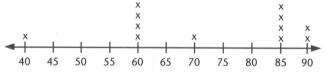

You start with a section of the number line that includes the test scores. Mark an **X** to record each score as it appears.

TRY THESE

1. In the box at the right is data collected about favorite flowers. Make a tally chart to organize this data: **R**–Rose, **D**–Daisy, **B**–Buttercup, and **L**–Lilly.

2. Organize the following temperatures using a line plot: 50, 50, 53, 53, 55, 57, 57, 57, 60, 60, 60, 60, and 60.

R	R	L	D	B
D	L	D	R	B
B	L	B	R	D
D	R	L	R	B

Exercises

1. Twelve students in Mr. Torres' science class have birthdays in the summer months: Mark (July), Samuel (August), Trina (June), Khesa (August), Morgan (July), Kelly (June), Rufus (June), Edward (July), Maria (July), Wendy (August), Wanda (June), and Felicia (June). Organize this birthday data into a tally chart.

2. Sasha's brownie troop had the following results in its cookie sales this year: Sasha (5 boxes), Mara (8 boxes), Natalie (3 boxes), Sonya (5 boxes), TuTu (7 boxes), Martha (10 boxes), Cecilia (3 boxes), Caitlin (5 boxes), and Vickie (2 boxes). Create a line plot for this data.

PROBLEM SOLVING

3. Organize the data in **Exercises 1** into a tally chart comparing the number of boys to girls having summer birthdays. Who has more summer birthdays—boys or girls? How many more?

4. Jacob rolled a die 24 times and got the following outcomes. Organize his results into a line plot. What do you notice?

   6 6 1 2 4 5 5 6 4 3 1 1
   2 4 4 3 3 2 5 2 3 3 3 1

5. Conduct a **survey**. In a survey, questions are asked and information is collected. Ask 10 people this question: Which is your favorite season: spring, summer, fall, or winter? Make a tally chart of your results.

MIND BUILDER

Using Patterns

Solve. Use the figure at the right.

1. If the figure is continued, how many letters will be in the **T** column?

2. What column will contain 15 letters?

```
              V
          W   V
        X   W   V
      Y   X   W   V
    Z   Y   X   W   V
      Y   X   W   V
        X   W   V
          W   V
              V
```

5.2 Mean, Median, Mode, and Range

Objective: to find the mean, median, mode, and range of a set of data

Pedro and Darius went to the zoo. They took a survey of the first twenty-three animals they saw and organized the data into the tally chart at the right.

Animal	Tally	Total Number Counted
Camel	||	2
Elephant	|||	3
Lion	||||	4
Hippopotamus	||||| ||	7
Giraffe	||||| ||	7

In order to describe their data, they used the range, median, mode, and mean.

The **range** is the difference between the greatest and the least number in a set of data.

Least number of animals seen: 2 camels
Greatest number of animals seen: 7 giraffes and 7 hippopotamuses
Range of animals seen: 7 – 2 = 5 animals

The **median** is the middle number when the data are listed in order.

2, 3, **4**, 7, 7
↑
median

The median of the total number of animals counted is 4.

The **mode** is the number that occurs most frequently.

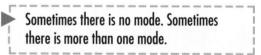

2, 3, 4, **7, 7**
most frequent = the mode

▶ Sometimes there is no mode. Sometimes there is more than one mode.

The mode of the zoo animal data is 7.

The **mean**, or **average**, is the sum of the numbers divided by the number of addends. To find the mean, a calculator is often a very helpful tool.

With the zoo animal data, it would look like this:
2 + 3 + 4 + 7 + 7 = 23 ÷ 5 animals. Mean = 4.6

Sometimes the mean can fall between whole numbers. Since you cannot see 4.6 animals, you would round to 5, the nearest whole number.

Another Example

40	45	45	45	45	60	75	90	95

The least number is 40.
The range is 95 – 40 = 55.
The mode (or most frequent number) is 45.
The mean (or average) is 60.

The greatest number is 95.
The median (or middle number) is 45.

$$40 + 45 + 45 + 45 + 45 + 60 + 75 + 90 + 95 = 540 ÷ 9 = 60$$

TRY THESE

Find the range, mode, median, and mean of each set of numbers.
(*Hint:* Put the numbers in numerical order first.)

1. 9, 6, 9

2. 3, 5, 5, 7, 10

3. 16, 23, 18

Exercises

Find the range, mean, median, and mode.

1. 8, 12, 6, 3, 6

2. 29, 29, 32

3. 24, 12, 12, 25, 27

4. 7, 1, 3, 5, 7, 9, 10

5. 5, 25, 15, 15, 10

★ 6. 27, 12, 7, 13, 5, 7, 8, 9, 20

PROBLEM SOLVING

Use the table at the right to solve.

Pounds of Food Eaten Weekly	
Camel	140
Elephant	1,215
Lion	72
Hippopotamus	518
Giraffe	280

7. Which animal eats the least amount of food?

8. Which animal eats the most amount of food?

9. What is the range of the least and most amounts of food?

10. How much food is needed to feed all 5 animals for a week?

CONSTRUCTED RESPONSE

Use the preceding table to solve.

11. Thomas argued that the hippopotamus eats more food weekly than the camel, lion, and giraffe combined. Is he right? Explain how you know.

12. Ellen says that the median of this data set is the lion's 72 pounds of food. Is she right? Explain how you know.

5.3 Problem-Solving Application: Using Statistics

Objective: to solve problems using the mean, median, and mode

Paulo's Plant Shop has four stacks of gardening books. He has 2 books in one stack, 5 in another, 2 in the third stack, and 3 in the fourth stack.

Paulo wants to arrange the books so that each stack has the same number of books. How many books will be in each of his stacks?

1. READ You know that there are different amounts of books in each stack. The picture and the list tell you how many are in each stack now. You know that Paulo wants equal stacks using these books.

2. PLAN Consider whether the mean, median, or mode will help you make equal stacks. The middle number of 2, 2, 3, and 5 (the median) will not help you. The number that is the most frequent (the mode) will not help you. Since you want all of the stacks to be equal, you want to find the mean. You can add together the four book stacks and use a calculator to divide.

3. SOLVE Follow your add-and-divide plan.

$$2 + 5 + 2 + 3 = 12$$

$$12 \div 4 = 3$$

Paulo can put 3 books in every stack to make equal stacks.

4. CHECK You can check by drawing a picture. When you draw a picture, you can see that 3 is a reasonable answer.

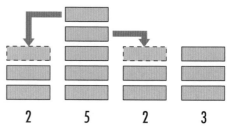

2   5   2   3

TRY THESE

Tell which measure would be the best to use to solve each situation: *mean*, *median*, or *mode*.

1. Harold surveys his class to find out what color is the most popular.

2. Sara wants to find the average price of tomatoes if they are 29¢ per pound on Monday, 31¢ per pound on Wednesday, and 33¢ per pound on Friday.

3. Mr. Rodriguez lines up his students by height to find who needs to be in the center of the stage for the performance.

> **REMEMBER:**
>
> **Mean** = the average of all items when added together and then divided by the total number of items
>
> **Mode** = the most frequent item
>
> **Median** = the middle item when put in order

Solve

1. The following are Nina's scores on 4 quizzes: 9, 7, 8, and 8. What is the mode of Nina's quiz scores?

2. Marsha ran 16 laps on Monday, 18 laps on Tuesday, and 14 laps on Wednesday. What is the mean number of the laps she ran?

3. A car was driven 13,378 miles the first year, 14,982 miles the second year, and 14,000 miles the third year. What is the average number of miles the car was driven in 3 years?

Use the calendar to solve problems 4–5.

4. What is the most frequent temperature in December?

5. Find December's median temperature.

December

Sun.	Mon.	Tues.	Wed.	Thurs.	Fri.	Sat.
1 45°F	2 42°F	3 38°F	4 37°F	5 37°F	6 32°F	7 30°F
8 28°F	9 25°F	10 27°F	11 32°F	12 30°F	13 32°F	14 31°F
15 26°F	16 30°F	17 21°F	18 19°F	19 15°F	20 14°F	21 18°F
22 22°F	23 29°F	24 30°F	25 30°F	26 25°F	27 21°F	28 16°F
29 13°F	30 12°F	31 14°F				

TEST PREP

6. Which data measure did you use to solve problem 4 above?

   a. mean   **b.** median   **c.** mode

7. At Northpoint Elementary School, the class sizes of five classrooms are 22, 31, 25, 18, and 24 students. Find the mean of this data set.

   a. 22   **b.** 25   **c.** 31   **d.** 24

5.3 Problem-Solving Application: Using Statistics   111

Problem Solving

Who, Which, and What Color?

Benny, Angie, and Kenny each own a different pet. Each of the pets is a different color. Using the following clues, find out which pet each child owns and what color it is.

Clues

1. Benny's pet is tan.

2. Angie does not have a dog.

3. Benny's best friend has a hamster.

4. Benny's best friend's pet is white.

5. Benny's best friend is Kenny.

6. Angie's next door neighbor has a dog.

7. Angie lives next door to Benny.

8. One of the three children has a brown cat.

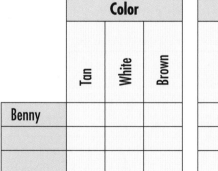

Copy and complete the grid. Write all names and categories in the grid. Use the clues to mark the boxes in the grid. Mark the boxes **X** for no and write *yes* for yes.

	Color			Pet		
	Tan	White	Brown			
Benny						

CUMULATIVE REVIEW

Complete each problem.

1. Write the word form of 9,567,123.

2. Which digit is in the thousands place in 246,801?

Solve.

3. 200
   × 5

4. 27
   × 3

5. 76
   × 2

6. 425
   × 4

7. 60
   × 3

8. 35
   × 7

9. 34
   × 5

10. 56
    × 8

11. 345
    × 2

12. 500
    × 6

13. 86
    × 1

14. 13
    × 9

15. Find the range of the following set of numbers: 11, 11, 24, 34, 56, and 99.

16. Find the mode of: 12, 12, 9, 6, 9, 5, and 12.

17. Find the mean of: 5, 5, 6, 6, and 8.

18. Find the median of: 6, 7, 8, 4, 7, 9, 10, 13, and 25.

Use the table to answer the following questions.

After Betsy walked through the woods, she decided to tally the number of different trees that she had seen.

19. Which type of tree did Betsy see the most?

20. Which type of tree did Betsy see the least?

21. What is the range of this data?

22. What is the mean of this data?

Types of Trees	Tally	Number of Trees
Spruce	ⅢⅠ	5
Cedar	ⅢⅠ	4
Maple	ⅢⅠ ⅢⅠ	8
Pine	ⅢⅠ ⅠⅠ	7

5.4 Single- and Double-Bar Graphs

Objective: to create and interpret single- and double-bar graphs

Dora and Tom each decided to survey their classmates about their favorite sport. Dora asked all of the girls, and Tom asked all of the boys. They combined their data into a shared table, and then they each made a **bar graph** of their own information.

Sports	Dora's Data: Girls	Tom's Data: Boys
Football	4	10
Soccer	5	5
Basketball	2	8
Hockey	3	9
Tennis	8	1

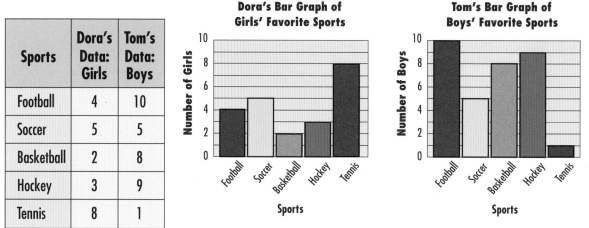

Their teacher challenged them to come up with a way to put both of their individual **single-bar graphs** onto a **double-bar graph** so the class could compare the two at the same time. Tom's and Dora's information is shown in the double-bar graph at the right.

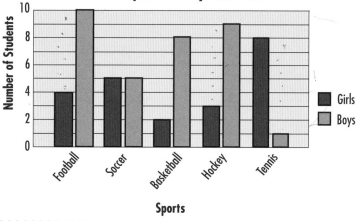

TRY THESE

Use the preceding double-bar graph to answer these questions.

1. Which sport do the boys in Tom's and Dora's class like the best? the least?

2. Which sport do the girls in Tom's and Dora's class like the best? the least?

3. Which sport do the boys and girls like equally?

4. Which one sport had the greatest difference between the girls and boys?

Exercises

Copy the graph to the right.

1. Make a bar graph showing the votes for favorite cold-climate animals. Use the data: penguin–3, polar bear–9, seal–12, and walrus–6.

2. How many total votes are shown?

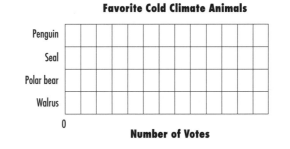

Favorite Cold Climate Animals

Penguin
Seal
Polar bear
Walrus

0

Number of Votes

PROBLEM SOLVING

Copy the chart to the right.

3. Organize the following data into the chart. **S** is a small fish; **M** is a medium fish; and **L** is a large fish.
   S M S S L M M S
   S S S L S M S S S

   Make a bar graph of the data.

4. Make a double-bar graph using the data in the chart on the right to show the number of hikers that visit each trail in each season.

Fish Harvest	
Size	Number of Fish Caught

Number of Hikers Per Trail

Trails	Summer	Winter
Valley Ridge	35	14
Eagle's View	41	22
Castle Falls	30	16
Winding Willows	27	18
Clayton's Cliff	19	8

TEST PREP

Use the double-bar graph you created in Problem Solving 4 to answer the following questions.

5. Which trail has the greatest number of summertime visitors?

   a. Winding Willows **b.** Castle Falls   **c.** Eagle's View   **d.** Valley Ridge

6. Which trail has the least number of visitors in the winter?

   a. Castle Falls   **b.** Valley Ridge   **c.** Eagle's View   **d.** Clayton's Cliff

5.5 Line Graphs

Objective: to create and interpret single-line graphs to show change

Members of the Animal Friends Club collect aluminum cans to recycle. They get paid 50¢ for every pound of cans that is recycled. Last Saturday, the members collected 12 pounds of aluminum cans. Club members want to show the value of the recycled aluminum cans. Using the data, they can find the value of the cans collected last Saturday.

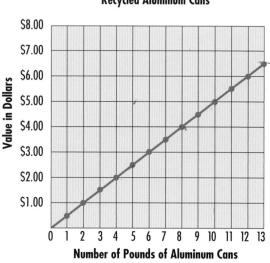

Instead of making bars to show your information, you can make a **line graph**. In a line graph, you plot your data on a graph using points, and then you connect the points of the line. Sometimes the line will be straight, as shown here, and sometimes it will go up and down between the points. Line graphs are very helpful to show change or to show a relationship.

In the preceding line graph, notice how the Animal Friends Club members labeled one scale, "Number of Pounds of Aluminum Cans" and labeled the other scale, "Value in Dollars." Each point on the graph is located by a number pair.

2 pounds	$1.00	(2, 1)
4 pounds	$2.00	(4, 2)
10 pounds	$5.00	(10, 5)
12 pounds	$6.00	(12, 6)

pounds of cans
↓
(12, 6)
↑
value in dollars

The number pair (12, 6) shows that the value of 12 pounds of aluminum cans is $6.00.

TRY THESE

What is the value of each amount of aluminum cans?
Use the graph on the previous page.

1. 6 pounds   **2.** 12 pounds   **3.** 13 pounds   **4.** 10 pounds

What amount of aluminum cans has each value?
Use the graph on the previous page.

5. $4.00   **6.** $5.00   **7.** $2.50   **8.** $6.50

Exercises

1. Copy and complete the line graph at the right to show the different amounts of snowfall for 6 weeks. Use the data in the chart.

Use the line graph you created in Exercises 1 to solve problems 2–3.

2. Which week had the most snowfall?

3. How many more inches of snow fell during week 1 than week 4?

6-Week Snowfall

Week	Amount
1	12 in.
2	8 in.
3	6 in.
4	5 in.
5	18 in.
6	10 in.

Title: _____

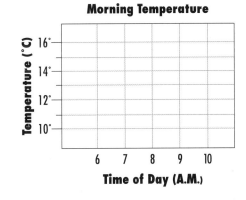

PROBLEM SOLVING

4. Copy and complete the line graph at the right. It shows the changes in morning temperature. Use this data: 6 A.M., 10°C; 7 A.M., 12°C; 8 A.M., 13°C; 9 A.M., 14°C; and 10 A.M., 16°C.

5. Using the preceding information, what is a good estimate for the temperature at 11 A.M.?

Morning Temperature

5.6 Stem-and-Leaf Plots

Objective: to create and interpret stem-and-leaf plots to show the distribution of data

Mrs. Strand's class just took a math test. There are 60 possible answers on the test. Mrs. Strand listed her students' math scores.

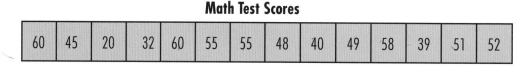

Math Test Scores

60	45	20	32	60	55	55	48	40	49	58	39	51	52

A **stem-and-leaf plot** can be used to help Mrs. Strand organize the scores. To begin, the data is listed from least to greatest. In a way, a stem-and-leaf plot is like a tree branch, where the stems can have many leaves. The ones digits are the **leaves**, and the tens digits are the **stems**.

Look at the parts of a stem-and-leaf plot. Keep these in mind when you make one of your own.

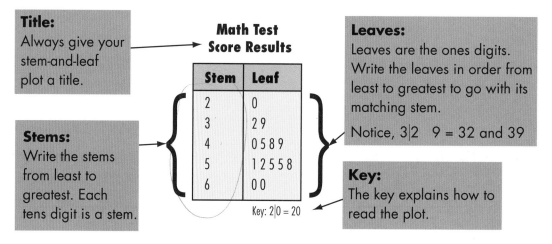

Title:
Always give your stem-and-leaf plot a title.

Math Test Score Results

Stem	Leaf
2	0
3	2 9
4	0 5 8 9
5	1 2 5 5 8
6	0 0

Key: 2|0 = 20

Stems:
Write the stems from least to greatest. Each tens digit is a stem.

Leaves:
Leaves are the ones digits. Write the leaves in order from least to greatest to go with its matching stem.

Notice, 3|2 9 = 32 and 39

Key:
The key explains how to read the plot.

TRY THESE

Use the preceding stem-and-leaf plot to answer these questions.

1. How many students are represented in the stem-and-leaf plot?

2. Find the median (the middle number) of the data.

3. Find the range for Mrs. Strand's test results.

4. In which group did the class have the most scores: twenties, thirties, forties, fifties, or sixties?

Exercises

Use the data to the right to complete problems 1–3.

1. Make a stem-and-leaf plot of the data.

2. What is the range from the oldest to youngest?

3. Find the average age of the people attending the reunion. What is another name for this measure?

Ages of People at the Family Reunion

25	12	11	10
5	32	45	41
53	17	42	67

PROBLEM SOLVING

Use the stem-and-leaf plot to the right to solve.

4. How many neighborhoods are represented in the data?

5. What is the range of speed limits among the neighborhoods of Trinity?

6. What does 3|5 mean in the stem-and-leaf plot?

7. Find the mean, median, and mode of the data set.

Neighborhood Speed Limits in Trinity, Florida

Stem	Leaf
1	0 0 5
2	0 5
3	0 0 5 5
4	0 0 0 5

Key: 1|0 = 10

CONSTRUCTED RESPONSE

8. Imagine that the additional speed limit of 55 was added to the preceding data. How would the stem-and-leaf plot change? How would the mean, median, and mode be affected?

MID-CHAPTER REVIEW

Use the bar graph to answer the following questions.

1. How many species of fish are endangered?

2. How many more species of birds than reptiles are endangered?

3. How many species of animals are shown altogether?

4. How many fewer species of insects than mammals are endangered?

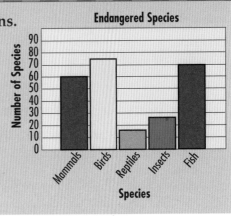

Endangered Species

5.7 Problem-Solving Application: Displaying Data

Objective: to decide the best graph to use to display data

The fourth grade students at Pinewood Elementary School collected data that is shown in the table at the right. They wanted to graph their data but were not sure which of the three graphs—pictograph, bar graph, or line graph—would be the best to use.

Mr. Korn, their teacher, divided the class into three groups, and each group was assigned a different graph to make from the data. Below are the pictograph, bar graph, and line graph made by the groups from the data. Together with Mr. Korn, the class decided to inspect the graphs closely to see which would be a better choice for their data.

Speeds of Animals	
Animals	Miles per Hour
Cheetah	70
Wildebeest	50
Lion	50
Elk	45
Zebra	40
Elephant	25

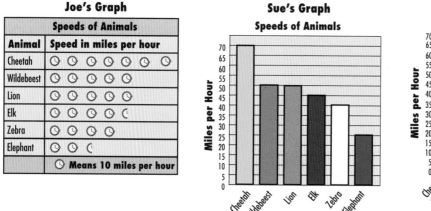

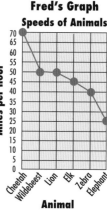

TRY THESE

Use the preceding graphs to answer each question.

1. What things are similar in the bar graph and the line graph?

2. How is the pictograph different from both the bar graph and the line graph?

Exercises

1. A line graph is best used when showing continuous change over a period of time.

   True or *False*: A line graph would be good to use for the students' data on the previous page.

2. Use your glossary to define these words:

   pictograph

   bar graph

PROBLEM SOLVING

Use the data on the previous page to solve.

3. Which animal has the fastest speed? Which animal has the slowest speed? What is the range of the given speeds?

4. What is the mean speed of the six animals?

5. Which two animals travel twice as fast as the elephant?

CONSTRUCTED RESPONSE

6. The class and Mr. Korn agreed that the bar graph on the previous page was the best way to display their data. Do you agree? Why or why not?

MIXED REVIEW

Write the number that is 1,000 more than each.

7. 2,437   8. 19,722   9. 209,999   10. 817,297

5.8 Coordinate Graphing

Objective: to graph points on a coordinate grid

Oliver stays in a hotel during the dog show. To locate Oliver's hotel room, start in the bottom left corner of the building. Count 3 windows to the right and 2 windows up.

3 units
to the right ——————┐ ┌—————— 2 units
up

(3, 2)

A pair of numbers called an **ordered pair** is used to locate Oliver's room. The same method can be used to locate a point on a grid. To locate a point, start at 0.

Another Example

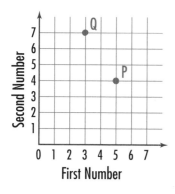

The ordered pair (5, 4) locates point P. Beginning at 0, count 5 units to the right and 4 units up.

5 units
to the right ——————┐ ┌—————— 4 units
up

(5, 4)

In an ordered pair, each individual number is called a **coordinate**. The first coordinate tells the number of units to the right of 0. The second coordinate tells the number of units up.

The ordered pair (3, 7) locates point Q.

TRY THESE

Write the ordered pair for each location.

1. Park   **2.** Theater   **3.** Bank   **4.** Library

Name the location of each ordered pair.

5. (1, 2)   **6.** (3, 2)   **7.** (2, 0)   **8.** (0, 0)

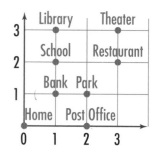

Exercises

Name the point for each ordered pair.

1. (1, 2)   **2.** (2, 5)   **3.** (6, 4)

4. (4, 6)   **5.** (5, 2)   **6.** (8, 5)

7. (7, 8)   **8.** (7, 0)   **9.** (0, 4)

Write the ordered pair for each point.

10. J   **11.** H   **12.** M   **13.** B

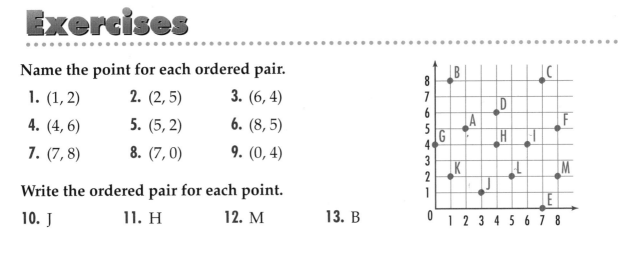

PROBLEM SOLVING

★ **14.** Plot each set of points on a new grid. Connect the points. Name the figures.

a. (5, 2), (3, 6), (1, 1)   **b.** (6, 0), (7, 3), (1, 5), (0, 2)

MIND BUILDER

Reading a Map

Use the map to answer the following questions.

1. What are the coordinates of Tampa?

2. Which city is located near (C, 11)?

3. Which is closer to (D, 5): St. Petersburg or the Sunshine Skyway Bridge?

4. Is Clearwater closer to (B, 7) or (A, 9)?

5. Near what point is Busch Gardens located?

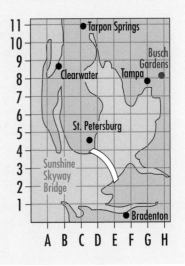

5.9 Function Tables

Objective: to complete function tables

Jacob has programmed his calculator to add 34 to each number that he enters. "Add 34" is the rule, or **function**, of his program.

Add 34.

Input	3	5	4	9	8
Output	37	39	38	43	42

The table he made is called a **function table**.

Jacob's teacher, Mrs. Bloom, showed him how he can use **variables** (letters or symbols) to write his function rule more clearly.

Using his teacher's suggestion, Jacob's function rule, "add 34," became $x + 34$.

His teacher also showed him another way to make his function table. Instead of placing the inputs and outputs in rows, she suggested placing them in columns. She also showed him how adding an extra column helps to avoid mistakes when determining the output. Mrs. Bloom said that the **input** is often known as the *x*-value, and the **output** is often known as the *y*-value.

In Mrs. Bloom's table, Jacob's data would look like this:

Input (x)	Function Rule x + 34	Output (y)
3	3 + 34	37
5	5 + 34	39
4	4 + 34	38
9	9 + 34	43
8	8 + 34	42

TRY THESE

Write these function rules using the variable *x*.

1. multiply by 7   **2.** add $3.50   **3.** subtract 17   **4.** divide by 2

Exercises

Copy and complete.

1.

Input (x)	Function Rule x × 4	Output (y)
2		
4		
6		
8		
10		

2.

Input (x)	Function Rule x ÷ 12	Output (y)
12		
24		
36		
60		

3.

Input (x)	Function Rule x − 11	Output (y)
22		
31		
44		
54		

4.

Input (x)	Function Rule (x + 4) − 2	Output (y)
6		
29		
43		
37		
72		

PROBLEM SOLVING

5. An online bookstore charges a $0.50 flat rate for postage and handling regardless of the number of $4.00 dictionaries bought. Copy and complete the chart at the right for this function.

6. What is the total cost for 6 dictionaries?

Input (x)	Function Rule (x × $4.00) + .50	Output (y)
1 dictionary	(x × $4.00) + .50	$4.50 total
2 dictionaries		total
3 dictionaries		total
4 dictionaries		total

5.10 Graphing Function Tables

Objective: to use function tables to graph a solution set

Ralph's Automotive needs to order more tires. Ralph knows that he usually replaces a complete set of tires for one car each week. How many tires will he need to order ahead for the next month?

You can use a function table to help Ralph figure out how many tires to buy.

Rule: $y = x \times 4$			
Input (x)	x × 4	Output (y)	Ordered Pair (x, y)
1 week	1 × 4	4 tires	(1, 4)
2 weeks	2 × 4	8 tires	(2, 8)
3 weeks	3 × 4	12 tires	(3, 12)
4 weeks (or 1 month)	4 × 4	16 tires	(4, 16)

You can also graph this information on a grid by using the table and your input (the *x*-value) and output (the *y*-value). You will have a line graph of your information when you finish.

1. Graph each point.

2. Connect the points to form a line.

This line is the graph of $y = x \times 4$.

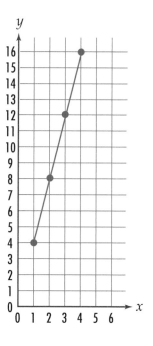

The horizontal number line on the grid is the **x-axis**. The vertical number line on the grid is the **y-axis**. The plural of axis is **axes**. The place where the *x*-axis meets the *y*-axis is called the **origin**. The coordinates of the origin are (0,0).

TRY THESE

Copy and complete. Then make the graph of each function.

1.

Rule: $y = x \times 4$			
Input (x)	$x \times 4$	Output (y)	Ordered Pair (x, y)
0 wagons	0×4	0 wheels	(0, 0)
1 wagon	1×4	__ wheels	(1, __)
2 wagons	2×4	__ wheels	(2, __)
3 wagons	3×4	__ wheels	(3, __)

2.

Rule: $y = x \div 2$			
Input (x)	$x \div 2$	Output (y)	Ordered Pair (x, y)
2 eyes	$2 \div 2$	1 person	(2, 1)
4 eyes	$4 \div 2$	__ people	(4, __)
6 eyes	$6 \div 2$	__ people	(__, __)
8 eyes	$8 \div 2$	__ people	(__, __)

Exercises

Copy and complete. Then make the graph of each function.

1.

Rule: $y = x + \$4$			
Input (x)	$x + 4$	Output (y)	Ordered Pair (x, y)
$1	$1 + 4$	__ total in bank	(__, __)
$2	$2 + 4$	__ total in bank	(__, __)
$3	$3 + 4$	__ total in bank	(__, __)
$4	$4 + 4$	__ total in bank	(__, __)

2.

Rule: $y = x \times 3$			
Input (x)	$x \times 3$	Output (y)	Ordered Pair (x, y)
0 triangles	0×3	0 corners	(0, 0)
1 triangle	1×3	__ corners	(__, __)
2 triangles	2×3	__ corners	(__, __)
3 triangles	3×3	__ corners	(__, __)

PROBLEM SOLVING

3. Delaney gets paid $5 per hour when she babysits. Use the function table to the right to determine how much money Delaney will make after babysitting for 2 hours, 4 hours, and 6 hours. Make a graph to show her rate of pay.

Rule: $x \times 5$	
Input (x)	Output (y)
2 hours	$ __
4 hours	$ __
6 hours	$ __

★ **4.** Dennis planted a daisy plant in his mom's garden. It measured 4 inches tall after he planted it. Every week it grew 2 inches. Write a function rule for the growth. Then graph the growth for its first 4 weeks in his mom's garden.

CHAPTER 5 REVIEW

LANGUAGE and CONCEPTS

Choose the correct word to complete each sentence.

1. The range of a set of numbers is the (sum, difference) of the least and greatest numbers.

2. The (mean, median) is the number that falls in the middle of an organized data set.

3. The mode is the (most frequent, average) number in a data set.

4. You can find the range, median, and mode from a (function table, stem-and-leaf plot).

5. When graphing points on a grid, the numbers in an ordered pair are called the (coordinates, inputs).

6. A function shows a (relationship, range) between the input and output numbers.

SKILLS and PROBLEM SOLVING

Find the range, mean, median, and mode for each set of numbers. (Section 5.2)

Range	Mean	Median	Mode

7. 9, 3, 6, 3, 4

8. 1, 2, 3, 1, 2, 2, 5, 2, 9

9. 17, 19, 19, 24, 28

Use the double-bar graph to solve. (Section 5.4)

10. What is the favorite forest animal in Class 4R?

11. What is the least favorite forest animal in Class 4J?

12. Which class has more students?

13. What is the range of choices for Class 4R?

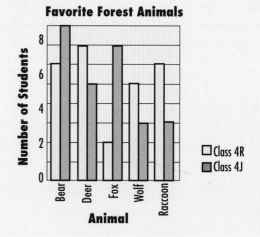

Favorite Forest Animals

Use the line graph on the right to solve. (Section 5.5)

14. Which season has the most vacationers?

15. What is the range of the data?

16. How many more people vacation in the winter than in the spring?

Use the following data to solve. (Section 5.6)

Heights of Students in Inches

61	58	49	52	59	60
48	52	53	62	58	58
49	55	57	54	56	55

17. Make a stem-and-leaf plot for the data

18. What stems did you need to make?

19. Which stem had the most leaves?

20. What is the mode of the data?

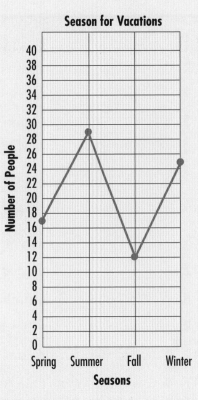

Season for Vacations

Graph the following ordered pairs on graph paper. (Section 5.8)

21. A(1, 5)   22. B(2, 2)   23. C(3, 4)   24. D(7, 3)

25. E(6, 3)   26. F(4, 4)   27. G(0, 5)   28. H(4, 3)

Copy and complete the function tables, then graph each function. (Sections 5.9–5.10)

29.

Rule: $12 – x	
Input (x)	**Output (y)**
$1 spent	$ ___ left
$3 spent	$ ___ left
$5 spent	$ ___ left
$9 spent	$ ___ left

30.

Rule: triple	
Input (x)	**Output (y)**
1 set	___ triplets
2 sets	___ triplets
3 sets	___ triplets
4 sets	___ triplets

31.

Rule: $x \div 2$	
Input (x)	**Output (y)**
6 socks	___ pairs of socks
10 socks	___ pairs of socks
12 socks	___ pairs of socks
4 socks	___ pairs of socks

CHAPTER 5 TEST

Use the chart at the right to solve.

1. Which animal has the slowest speed? What is its speed?

2. What is the range for these speeds?

3. What is the most frequent speed?

4. What is the median speed?

5. What is the average speed of all the animals?

6. Make a stem-and-leaf plot of this data.

Maximum Speeds of Animals	
Animal	Miles per Hour
Cheetah	70
Antelope	64
Lion	50
Gazelle	50
Coyote	44
Zebra	40
Giraffe	32
Grizzly bear	30
Elephant	25

Use the bar graph to answer each question.

7. How many push-ups does Laura do?

8. Who does the least number of push-ups?

9. Who does less than 10 push-ups?

10. How many push-ups does Shannon do?

11. You could use this data to make a line graph, but it is not the best type of graph for this data. Why?

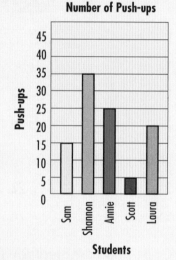

Graph the following ordered pairs on graph paper.

12. J(5, 1)

13. K(5, 5)

14. L(4, 3)

15. M(3, 7)

Copy and complete each function table, and then graph each function.

16.

Rule: x ÷ 5	
Input (x)	Output (y)
25 cents	____ nickels
5 cents	____ nickels
45 cents	____ nickels
10 cents	____ nickels

17.

Rule: x × 4	
Input (x)	Output (y)
1 tiger	____ legs
2 tigers	____ legs
3 tigers	____ legs
4 tigers	____ legs

18. Pizza World is running a grand-opening sale. For every 5 pizzas you order, you receive another 1 free. Create a function table to decide how many total pizzas you will receive if you order 10, 15, and 20 pizzas.

Circle Graphs

The monthly income of the Pinkard family is budgeted very carefully. The following circle graph displays the categories of the family's spending.

Monthly Family Budget

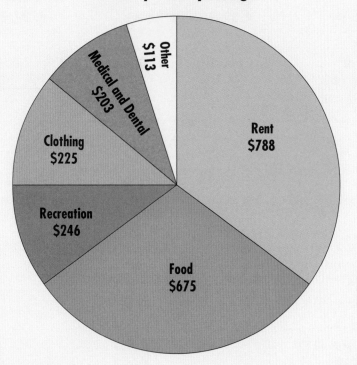

Circle graphs compare parts to a whole. Because of their shape, they sometimes are called a pie graph.

Study the circle graph and answer the following questions.

1. What is the family's total monthly income?

2. On what is most of the family's income spent?

3. Order the categories from least to greatest.

4. Give at least three possible items that might be classified as *Other*.

5. What is the range of the family's spending?

6. What is the family's annual income?

CUMULATIVE TEST

1. What is the missing addend?

   $\blacksquare + 7 = 9$

   a. 2
   b. 3
   c. 16
   d. none of the above

2.
   $$\begin{array}{r} \$48.65 \\ + \ 23.98 \\ \hline \end{array}$$

   a. $24.67
   b. $72.63
   c. $65.53
   d. none of the above

3.
   $$\begin{array}{r} 20,000 \\ - \ 9,851 \\ \hline \end{array}$$

   a. 10,149
   b. 11,259
   c. 13,559
   d. 29,851

4. Choose the pair of numbers that has one even number and one odd number.

   a. 4, 9
   b. 5, 7
   c. 6, 8
   d. none of the above

5. Choose the best estimate.
   $$\begin{array}{r} 689 \\ \times \ 7 \\ \hline \end{array}$$

   a. 4,200
   b. 4,900
   c. 10,000
   d. none of the above

6.
   $$\begin{array}{r} 7,046 \\ \times \ 5 \\ \hline \end{array}$$

   a. 35,030
   b. 35,150
   c. 35,230
   d. none of the above

7. Lois spent $18 for a skirt and $15 for a blouse. How can you find the total amount she spent?

   a. Subtract $15 from $18.
   b. Add $18 and $15.
   c. Compare $15 and $18.
   d. none of the above

8. Each warm-up suit costs $47.95. How much do 8 warm-up suits cost?

   a. $373.20
   b. $373.60
   c. $383.20
   d. none of the above

9. Rex has nine shirts. Four of his shirts are red, two are blue, and the rest are white. How many of his shirts are white?

   a. 3
   b. 11
   c. 15
   d. none of the above

10. How high was the corn plant at the end of the third week?

    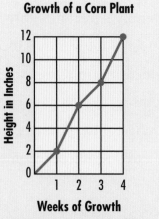

    Growth of a Corn Plant

    a. 2 inches
    b. 6 inches
    c. 8 inches
    d. 12 inches

Dividing by One-Digit Numbers

Matthew Cahn
Calvert Day School

6.1 Related Facts

Objective: to relate basic multiplication and division facts using inverse operations

Riding a camel is one way to travel in some parts of the world. Imagine looking down and seeing 32 camel legs. How many camels would that be?

Knowing that a camel has 4 legs, you want to know how many camels have a total of 32 legs.

$32 \div 4 = \blacksquare$

$4 \times 8 = 32$

THINK
To find the answer, think of multiplication facts for 4.

If there are 32 camel legs, there must be 8 camels.

More Examples

A. You can write four different multiplication and division facts using the numbers 4, 8, and 32. This is called a **fact family**.

$4 \times 8 = 32$ $8 \times 4 = 32$

$32 \div 8 = 4$ $32 \div 4 = 8$

B. You can use a multiplication table to find related facts and fact families.

$63 \div 9 = \blacksquare$ $9 \times \blacksquare = 63$

Look in the 9 row for 63. Then look up to find the other factor, 7.

TRY THESE

Match each related fact.

1. $6 \times 6 = 36$
2. $36 \div 9 = 4$
3. $3 \times 8 = 24$
4. $16 \div 4 = 4$
5. $24 \div 6 = 4$
6. $1 \times 9 = 9$

a. $6 \times 4 = 24$
b. $9 \div 9 = 1$
c. $24 \div 8 = 3$
d. $4 \times 9 = 36$
e. $4 \times 4 = 16$
f. $36 \div 6 = 6$

×	0	1	2	3	4	5	6	7	8	9
0	0	0	0	0	0	0	0	0	0	0
1	0	1	2	3	4	5	6	7	8	9
2	0	2	4	6	8	10	12	14	16	18
3	0	3	6	9	12	15	18	21	24	27
4	0	4	8	12	16	20	24	28	32	36
5	0	5	10	15	20	25	30	35	40	45
6	0	6	12	18	24	30	36	42	48	54
7	0	7	14	21	28	35	42	49	56	63
8	0	8	16	24	32	40	48	56	64	72
9	0	9	18	27	36	45	54	63	72	81

Exercises

Write a related multiplication fact.

1. 40 ÷ 8 = 5   **2.** 56 ÷ 7 = 8   **3.** 32 ÷ 4 = 8   **4.** 72 ÷ 8 = 9

Write each fact family.

5. 3, 6, 18   **6.** 8, 5, 40   **7.** 6, 9, 54   **8.** 8, 6, 48   **9.** 9, 8, 72

10. 42, 6, 7   **11.** 45, 9, 5   **12.** 12, 132, 11   **13.** 16, 240, 15   **14.** 420, 21, 20

PROBLEM SOLVING

15. The product of 3 and 4 is 12. Name another pair of numbers whose product is 12.

16. The product of 1 and 24 is 24. Name two other pairs of numbers whose product is 24.

17. The numbers 8 and 2 belong to a fact family. Name two numbers that could be the third member of this fact family.

18. Sean's mom gives him a big bowl of 35 strawberries. She asks Sean to put the same amount of strawberries in 5 bowls for the family's dessert. How many strawberries will be in each bowl?

MIND BUILDER

Fact Triangles

Fact triangles are useful for helping you see fact families. The starred number in the top corner is the product of the bottom two corners. You can find the related division fact as well.

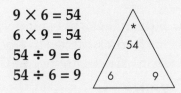

$9 \times 6 = 54$
$6 \times 9 = 54$
$54 \div 9 = 6$
$54 \div 6 = 9$

Study the following fact triangles to solve for the missing number. Write the fact family.

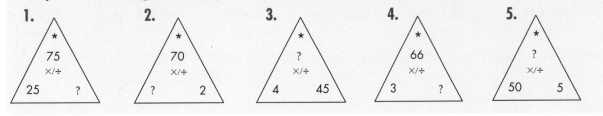

1.
*
75
×/÷
25 ?

2.
*
70
×/÷
? 2

3.
*
?
×/÷
4 45

4.
*
66
×/÷
3 ?

5.
*
?
×/÷
50 5

6.2 Missing Factors

Objective: to solve for missing factors by using division

Fifteen vehicles are arranged evenly in three rows. How many are in each row? You need to divide 15 by 3.

Division is the operation used to separate a group into smaller groups of the same number. It is also used to find how many *equal* groups can be formed.

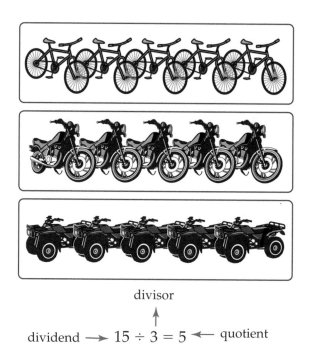

total number		number of groups		number in each group
15	÷	3	=	■

THINK $3 \times ■ = 15$

15	÷	3	=	5
vehicles		rows		vehicles in each row

There are five vehicles in each row.

divisor

dividend ⟶ $15 ÷ 3 = 5$ ⟵ quotient

Say: Fifteen divided by three is equal to five.

More Examples

Division is also used to find the number of groups.

A. Fifteen boats are shipped to a store. Each truck carries five boats. How many trucks are needed?

THINK $■ \times 5 = 15$

total number	number in each group	number of groups
15	÷	5	=	3

B. Twelve skateboards are on display. Each row has three skateboards. How many rows are there?

THINK $■ \times 3 = 12$

total number	number in each group	number of groups
12	÷	3	=	4

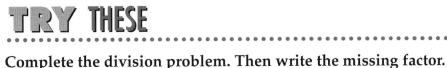

TRY THESE

Complete the division problem. Then write the missing factor.

1. $12 \div 2 = \blacksquare$

$2 \times \blacksquare = 12$

2. $27 \div 3 = \blacksquare$

$3 \times \blacksquare = 27$

3. $18 \div 6 = \blacksquare$

$6 \times \blacksquare = 18$

Exercises

Complete the division problem. Then write the missing factor.

1. $24 \div 6 = \blacksquare$

$6 \times \blacksquare = 24$

2. $5 \div 5 = \blacksquare$

$5 \times \blacksquare = 5$

3. $35 \div 7 = \blacksquare$

$7 \times \blacksquare = 35$

Complete the division problem. Then write the multiplication fact that helped you.

4. $72 \div 8 = \blacksquare$

5. $49 \div 7 = \blacksquare$

6. $63 \div 9 = \blacksquare$

7. $56 \div 7 = \blacksquare$

8. $36 \div 3 = \blacksquare$

9. $45 \div 5 = \blacksquare$

Complete.

★**10.** $4 \times 4 = 48 \div \blacksquare$

★**11.** $35 \div 5 = 7 \times \blacksquare$

★**12.** $36 \div 4 = \blacksquare \div 7$

PROBLEM SOLVING

13. Florence shared 21 motorcycle photos equally with 3 friends. How many photos did each friend receive?

14. Mrs. Smitson expects to sell 12 cars before Friday. She has sold 4 cars. How many cars does she need to sell?

CONSTRUCTED RESPONSE

15. There are 24 students in a class. For a field trip, the class is divided into groups. Each group has 6 students. How many groups are on the field trip? Explain how you determined your answer.

16. Three boys worked together for an hour picking strawberries. The farm owner gave them $27 to share equally among themselves. Was this a fair payment? Explain why you think it is or is not.

6.2 Missing Factors   **137**

6.3 Division Patterns

Objective: to divide mentally when the dividend is a multiple of 10

The school bought 80 bushes to plant in the front and the back of the school. If the same amount of bushes was planted in both places, how many bushes were planted in each place?

THINK What basic fact helps to solve this? How many 2s are in 80?

Compare the following problems.

$2\overline{)8}$ $2 \times \blacksquare = 8$ $2\overline{)8}^{4}$ $2\overline{)80}$ $2 \times \blacksquare = 80$ $2\overline{)80}^{40}$

How are the problems alike? How are the problems different? What pattern do you see?

Describe the pattern you would use to help solve this problem. $2\overline{)800}$

More Examples

A. $3\overline{)6}^{2}$ $3\overline{)60}^{20}$ $3\overline{)600}^{200}$

$3 \times \blacksquare = 6$

B. $18 \div 6 = 3$ $6 \times \blacksquare = 18$

$180 \div 6 = 30$

$1,800 \div 6 = 300$

TRY THESE

Divide mentally.

1. $7\overline{)7}$ $7\overline{)700}$ $7\overline{)7,000}$ **2.** $5\overline{)25}$ $5\overline{)250}$ $5\overline{)2,500}$

3. $9 \div 3 = \blacksquare$ **4.** $35 \div 5 = \blacksquare$ **5.** $40 \div 8 = \blacksquare$

$90 \div 3 = \blacksquare$ $350 \div 5 = \blacksquare$ $400 \div 8 = \blacksquare$

$900 \div 3 = \blacksquare$ $3,500 \div 5 = \blacksquare$ $4,000 \div 8 = \blacksquare$

Exercises

Divide mentally.

1. $5\overline{)5}$
2. $5\overline{)50}$
3. $5\overline{)500}$
4. $5\overline{)5,000}$

5. $5\overline{)50,000}$
6. $3\overline{)6}$
7. $3\overline{)60}$
8. $3\overline{)600}$

9. $3\overline{)6,000}$
10. $3\overline{)60,000}$
11. $9\overline{)81}$
12. $9\overline{)810}$

13. $9\overline{)8,100}$
14. $9\overline{)81,000}$
15. $9\overline{)810,000}$
16. $9\overline{)270}$

17. $6\overline{)300}$
18. $8\overline{)56,000}$
19. $4\overline{)2,800}$
20. $5\overline{)4,000}$

21. Norman planted 100 pumpkin seeds. He planted 5 seeds in each row. How many rows of pumpkin seeds did he plant?

22. Stephanie put 120 ornaments into 3 boxes. She made sure each box contained the same number of ornaments. How many ornaments were in each box?

23. Dave bought 16 onion plants, 12 tomato plants, and 24 lettuce plants. How many plants did he buy in all?

24. Sharon planted 210 strawberry plants in 7 rows in her garden. How many plants were in each row?

CONSTRUCTED RESPONSE

25. After looking at the division patterns, George told Paige, "There will always be at least the same number of zeros in the quotient as in the dividend." Paige disagreed and said, "Not always." Who is right? Explain your reasoning.

MIXED REVIEW

Find the product.

26. 634×87
27. 145×16
28. $85,098 \times 0$
29. 387×45

6.4 Remainders and Checking

Objective: to understand the meaning of remainders in
division problems

Mrs. Lee has a pack of 26 jelly beans. She wants to divide them equally among her 3 children. How many will each child receive? Will there be any jelly beans left over?

To find out how many jelly beans each of her children will receive, divide.

Step 1	Step 2	Step 3	Step 4	Step 5
Divide.	Multiply.	Subtract.	Write the remainder beside the **quotient**.	Check by multiplication and addition.
$\begin{array}{r} 8 \\ 3\overline{)26} \end{array}$	$\begin{array}{r} 8 \\ 3\overline{)26} \\ 24 \end{array}$	$\begin{array}{r} 8 \\ 3\overline{)26} \\ -24 \\ \hline 2 \end{array}$	$\begin{array}{r} 8\,R2 \\ 3\overline{)26} \\ -24 \\ \hline 2 \end{array}$	$\begin{array}{r} 8 \\ \times\ 3 \\ \hline 24 \end{array}$ $\begin{array}{r} 24 \\ +2 \\ \hline 26 \end{array}$
$26 \div 3 = \blacksquare$ THINK $3 \times \blacksquare = 26$ There is no number fact for 3 to get 26. The closest number fact is $3 \times 8 = 24$. Write 8 in the ones place.	Multiply 3×8 and place the product 24 beneath the 26.	Subtract 24 from 26. The difference is called the **remainder**.	Use a capital **R** to show that all the items were not able to be divided into equal groups and there were some left over.	Does 3 times 8 plus 2 equal 26? Yes, $3 \times 8 = 24$, and $24 + 2 = 26$.

Each of Mrs. Lee's children will receive 8 jelly beans. There will be 2 jelly beans left over.

Another Example

A group of 9 people are waiting in line to get on boats. Each boat safely seats 5 people. How many boats are needed?

$\begin{array}{r} 1\,R4 \\ 5\overline{)9} \\ -5 \\ \hline 4 \end{array}$ To find the number of boats, divide 9 by 5.

The answer seems to be 1 with a remainder of 4, but this would not answer the question of how many boats are needed. For everyone in the group to be in boats, the answer that makes sense in this case is 2 boats.

TRY THESE

Divide.

1. 2)7   **2.** 4)23   **3.** 8)16   **4.** 5)47   **5.** 9)34   **6.** 7)20

Exercises

Find the quotient and the remainder. Check your answer by multiplying and adding.

1. 4)30   **2.** 4)19   **3.** 4)11   **4.** 4)16   **5.** 4)13   **6.** 5)48

7. 3)29   **8.** 7)66   **9.** 4)25   **10.** 3)20   **11.** 9)83   **12.** 6)39

13. 8)79   **14.** 9)59   **15.** 5)36   **16.** 8)26   **17.** 7)63   **18.** 5)39

PROBLEM SOLVING

19. Shenice bought 4 feet of string. Her bill was 36¢. What was the cost for each foot?

20. What is the cost of one hamster if three hamsters cost $30?

Use the chart to solve.

21. In all, how much will Abbie need to spend if she needs 4 inches of each trim?

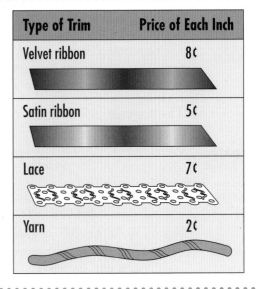

Type of Trim	Price of Each Inch
Velvet ribbon	8¢
Satin ribbon	5¢
Lace	7¢
Yarn	2¢

TEST PREP

22. Ian needs to purchase 17 soccer balls. They cost $4.95 each. How much will Ian spend?

    a. $56.15   **b.** $84.15   **c.** $67.35   **d.** $82.15

6.5 Problem-Solving Application: Interpret Remainders

Objective: to interpret remainders properly when solving problems

When you divide to solve some word problems, your answer may have a remainder. You must decide what to do with the remainder. Study these two problems.

A

1. READ | A store sells 5-pound bags of seed. How many bags can be filled from 38 pounds of seed?

2. PLAN | To find out how many 5 pounds are in 38 pounds, divide.

3. SOLVE

$$
\begin{array}{r}
7 \text{ R3} \\
5\overline{)38} \\
-35 \\
\hline
3
\end{array}
$$

If 7 bags are filled, there are 3 pounds left.

The remaining 3 pounds will not fill a bag. You cannot use the remainder to answer the question.

The answer is 7 bags.

4. CHECK

$5 \times 7 = 35$ 7 bags, 35 pounds

$5 \times 8 = 40$ 8 bags, 40 pounds

There are 38 pounds of seed. Choose the product that is a little less than 38. So, 7 bags can be filled.

B

Mr. Sims needs 17 pounds of seed. How many 5-pound bags should he buy?

To find out how many 5-pound bags are in 17 pounds, divide.

$$
\begin{array}{r}
3 \text{ R2} \\
5\overline{)17} \\
-15 \\
\hline
2
\end{array}
$$

If he buys 3 bags, he will need 2 more pounds.

To get 2 more pounds, Mr. Sims needs 1 more bag.

The answer is 4 bags.

$5 \times 3 = 15$ 3 bags, 15 pounds

$5 \times 4 = 20$ 4 bags, 20 pounds

Mr. Sims needs 17 pounds or more. Choose the product that is a little more than 17. So, Mr. Sims should buy 4 bags.

TRY THESE
. . . .

Solve.

1. Mrs. Kim has $23 to spend on flower pots. If each pot costs $7, how many can she buy?

2. Mr. Thomas needs 42 pounds of fertilizer. How many 5-pound bags should he buy?

Solve

1. Rick has 83 pepper plants to put on trays. He can put 9 plants on each tray. How many trays does he need?

2. The school spent $600 for new trays in the lunchroom. Each tray cost $6. How many trays did the school buy?

3. A pound of dry fertilizer makes 6 gallons of spray. How many pounds are needed to make 49 gallons of spray?

4. Robert needs 65 tomato stakes. The stakes are sold in bundles of 8. How many bundles does he need?

5. Bev paid $2.98 for a plant and $4.99 for a planter. She gave the clerk $10.00. How much change did Bev receive?

6. Trisha wants to build a fence 74 feet long. How many 8-foot sections does she need?

7. Garnet bought 20 feet of rope. How many 3-foot pieces can she make?

8. Keith has $37. He wants to buy some books that cost $7 each. How many books can he buy? How much money will he have left over?

9. Twenty-eight students are going to the art museum. Five students can go in each car. How many cars are needed?

10. Owen's grandmother has 38 tulip bulbs. She wants to plant them in rows of 6. How many rows can she plant?

11. Emma, Jill, and Ed are moving. They can pack 8 books in each carton. How many cartons do they need for 75 books?

★12. Sara's father bought party hats for $1.12. Each hat cost 8¢. How many hats did he buy?

CONSTRUCTED RESPONSE

13. Molly bought a bag of chocolates to share with 8 of her friends. There are 60 chocolates per bag. How many chocolates will each girl receive? Will Molly have any extra chocolates to share with her teacher? Explain your reasoning.

14. Jane has 15 quarters. Bob has 75 nickels. Who has more money? How can you tell?

6.6 Dividing with No Regrouping

Objective: to divide two-digit numbers with no regrouping

A street has 86 trees along the sides. Each side has the same number of trees. How many trees are along each side? To find the answer, divide 86 by 2.

Step 1	Step 2	Step 3
Write the problem. $2\overline{)86}$ Think of the multiplication facts for 2.	Divide the tens. tens : ones $\begin{array}{r} 4 \\ 2\overline{)8\,6} \\ -8 \\ \hline 0 \end{array}$ Divide. $8 \div 2 = \blacksquare$ Multiply. $2 \times 4 = 8$ Subtract. $8 - 8 = 0$ Compare. The remainder must be less than the divisor.	Divide the ones. tens : ones $\begin{array}{r} 4\,3 \\ 2\overline{)8\,6} \\ -8\downarrow \\ \hline 0\,6 \\ -6 \\ \hline 0 \end{array}$ Bring down the 6. Divide. $6 \div 2 = \blacksquare$ Multiply. $2 \times 3 = 6$ Subtract. $6 - 6 = 0$

There are 43 trees along each side of the street.

More Examples

A.
$\begin{array}{r} 21 \\ 4\overline{)84} \\ -8\downarrow \\ \hline 04 \\ -4 \\ \hline 0 \end{array}$
  Divide the tens.
  $4 \times \blacksquare = 8$
  Divide the ones.
  $4 \times \blacksquare = 4$

B.
$\begin{array}{r} 22 \text{ R } 1 \\ 3\overline{)67} \\ -6\downarrow \\ \hline 07 \\ -6 \\ \hline 1 \end{array}$
  The remainder must be less than the divisor.

C.
$\begin{array}{r} \$11 \text{ R}\$2 \\ 4\overline{)\$46} \\ -4\downarrow \\ \hline 06 \\ -4 \\ \hline 2 \end{array}$
  Write the $ sign in the quotient and remainder.

TRY THESE

Divide.

1. $2\overline{)62}$ 2. $7\overline{)77}$ 3. $2\overline{)48}$ 4. $3\overline{)69}$ 5. $6\overline{)66}$ 6. $5\overline{)\$57}$

Exercises

Divide.

1. $2\overline{)28}$ **2.** $4\overline{)87}$ **3.** $2\overline{)46}$ **4.** $8\overline{)88}$ **5.** $6\overline{)\$69}$

6. $3\overline{)37}$ **7.** $3\overline{)96}$ **8.** $4\overline{)\$47}$ **9.** $2\overline{)66}$ **10.** $3\overline{)35}$

11. $3\overline{)98}$ **12.** $2\overline{)\$67}$ **13.** $4\overline{)86}$ **14.** $3\overline{)39}$ **15.** $2\overline{)89}$

16. $67 \div 3$ **17.** $88 \div 4$ **18.** $35 \div 3$ **19.** $29 \div 2$ **20.** $\$24 \div 2$

21. $\$34 \div 3$ **22.** $\$62 \div 2$ **23.** $\$43 \div 2$

Solve.

⭐ **24.** Find the quotient for $49 \div 2$.

⭐ **25.** Compute. $(48 \div 2) - 9 =$ _____

MID-CHAPTER REVIEW

Divide.

1. $5\overline{)25}$ **2.** $5\overline{)250}$ **3.** $5\overline{)2,500}$ **4.** $5\overline{)54}$ **5.** $8\overline{)89}$

6. $7\overline{)\$73}$ **7.** $4\overline{)47}$ **8.** $3\overline{)97}$ **9.** $4\overline{)89}$ **10.** $2\overline{)89}$

Solve.

11. Find the price of 1 can using the following information. Each can costs the same.

12. The length of a hallway is 39 feet. What is the length in yards? (*Hint:* 1 yard is 3 feet.)

13. A box holds 8 crayons. How many crayons are in 64 boxes?

14. Michelle has 58 lettuce seeds. She plants them evenly in 5 rows. How many seeds are in each row? How many are left over?

6.7 Dividing with Regrouping

Objective: to divide two-digit numbers by one-digit numbers when regrouping is needed

Beth, Amy, and Mark picked 48 watermelons. They each picked the same number. How many watermelons did each person pick? To find the answer, divide 48 by 3.

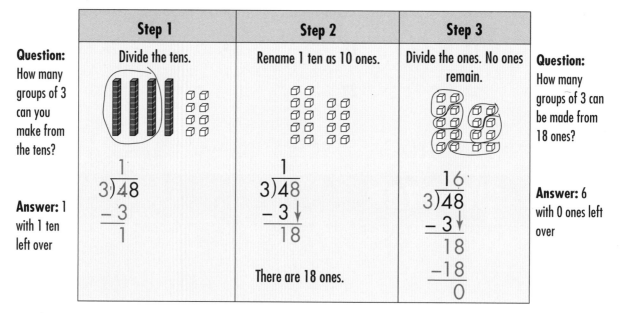

	Step 1	**Step 2**	**Step 3**	
Question: How many groups of 3 can you make from the tens?	Divide the tens.	Rename 1 ten as 10 ones.	Divide the ones. No ones remain.	**Question:** How many groups of 3 can be made from 18 ones?
Answer: 1 with 1 ten left over	$\begin{array}{r} 1 \\ 3\overline{)48} \\ -3 \\ \hline 1 \end{array}$	$\begin{array}{r} 1 \\ 3\overline{)48} \\ -3\downarrow \\ \hline 18 \end{array}$ There are 18 ones.	$\begin{array}{r} 16 \\ 3\overline{)48} \\ -3\downarrow \\ \hline 18 \\ -18 \\ \hline 0 \end{array}$	**Answer:** 6 with 0 ones left over

Each person picked 16 watermelons.

The steps in division are Divide, Multiply, Subtract, Compare, and Bring Down. In order to help you remember the steps, try to come up with a fun sentence like the example on the right:

Does
Mark
Sell
Cheese
Burgers

More Examples

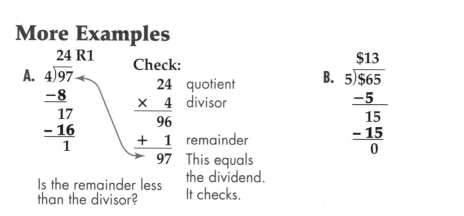

A.
$\begin{array}{r} 24\text{ R1} \\ 4\overline{)97} \\ -8 \\ \hline 17 \\ -16 \\ \hline 1 \end{array}$

Is the remainder less than the divisor?

Check:
$\begin{array}{r} 24 \\ \times\ \ 4 \\ \hline 96 \\ +\ \ 1 \\ \hline 97 \end{array}$ quotient
divisor

remainder
This equals the dividend. It checks.

B.
$\begin{array}{r} \$13 \\ 5\overline{)\$65} \\ -5 \\ \hline 15 \\ -15 \\ \hline 0 \end{array}$

TRY THESE

Find the quotient. Remember to use the steps divide, multiply, subtract, compare, and bring down (regroup).

1. 4)92 **2.** 2)74 **3.** 3)$81 **4.** 2)52 **5.** 4)47 **6.** 5)63

Exercises

Divide.

1. 3)45 **2.** 6)78 **3.** 4)67 **4.** 7)85 **5.** 3)87 **6.** 5)96

7. 4)95 **8.** 2)35 **9.** 7)91 **10.** 6)90 **11.** 2)73 **12.** 3)77

13. 6)$84 **14.** 2)$51 **15.** 5)$60 **16.** 4)$95 **17.** 7)$79 **18.** 4)$72

PROBLEM SOLVING

19. Rachel is making plant hangers. She needs 4 beads for each hanger. She has 49 beads. How many hangers can she make? How many beads will be left?

20. Find the price of one orange.

6 for 96¢ Oranges

★ **21.** Miguel made 42 deliveries for Karl's Pizza Palace. Dudley made 7 deliveries for Karl's Pizza Palace. How many more deliveries did Miguel make?

★ **22.** Start with 168. Guess how many times you can subtract 7 to get to 0. Then try it. Can you think of a faster way to find out how many 7s are in 168?

```
  168
−   7
  161
−   7
  154
−   7
```

MIXED REVIEW

Find the product.

23. $19.99 × 23 **24.** 365 × 73 **25.** $48.73 × 41 **26.** 416,432 × 92

Problem Solving

Max's Travels

Max works for a delivery service. He delivers packages to 7 different cities each week. He can go from any city to any city by a direct route. How many different highways could Max travel in 1 week?

Trace this map on your paper to find the highways.

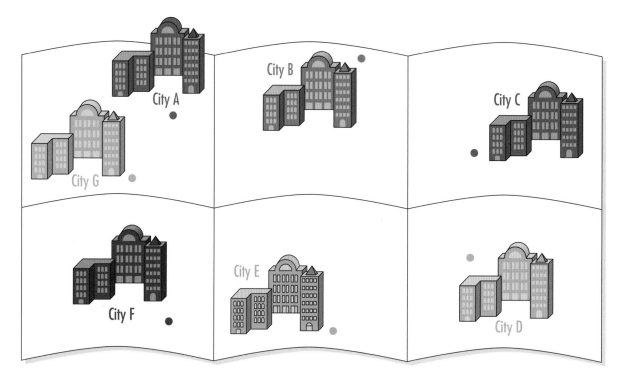

Extension

1. The delivery service expands its delivery to 3 more cities. On how many different highways could Max travel in 1 week?

2. The company decides to use City F as a hub. This means that all packages are first shipped to City F. From City F, Max delivers packages to the other cities. He can only travel to and from City F. He cannot go from any city to any city by a direct route. On how many different highways could Max travel in 1 week?

Write the number named by each 8.

1. 28,350
2. 458
3. 6,384
4. 822,637,000

Order the numbers in each list from least to greatest.

5. 11 8 24
6. 97 79 99 109
7. 500 498 489

Simplify.

8. $(57 + 26) - 15$
9. $(982 - 87) + 215$
10. $85 - (47 + 4)$

Compute.

11. $\begin{array}{r} 3,752 \\ -\ 1,231 \end{array}$
12. $\begin{array}{r} \$78.05 \\ -\ 19.37 \end{array}$
13. $\begin{array}{r} 57,892 \\ +\ 13,257 \end{array}$
14. $\begin{array}{r} 28,098 \\ -\ 19,258 \end{array}$
15. $\begin{array}{r} \$8.36 \\ +\ 0.53 \end{array}$

16. 3×4
17. 8×8
18. 25×9
19. 234×7
20. $\$573 \times 100$
21. $23 \times \$40$
22. 56×74
23. 37×56
24. 63×45
25. 92×35
26. 568×78
27. 320×87
28. 657×28
29. $1,364 \times 57$
30. $3,096 \times 42$
31. $2,851 \times 64$

32. $8\overline{)16}$
33. $3\overline{)27}$
34. $6\overline{)27}$
35. $2\overline{)42}$
36. $3\overline{)89}$
37. $5\overline{)73}$

Solve.

38. Jason earns $32 for working 8 hours. How much does Jason make an hour?

39. Sixty-seven people are on a bus. The bus limit has been reached with four people in each row plus 3 people standing. How many rows of seats does the bus have?

40. Danny and Della each have a bag of jelly beans. Danny has 7 red jelly beans, 5 green ones, 9 yellow ones, and 3 purple ones. Della has 8 red jelly beans, 3 purple ones, 5 yellow ones, and 7 green ones. Make a double-bar graph on graph paper using this information. Then write two or more sentences comparing Danny's and Della's jelly bean data.

Jelly Beans

6.8 Estimating Quotients

Objective: to use compatible numbers to estimate quotients

Each of the 543 students in Edgewood School will be given a tree to plant on Arbor Day. There are 6 grades in the school. About how many trees are needed for each grade?

> ▶ You can estimate a quotient if you know the following:
> - the number of digits in the quotient
> - the leading digit of the quotient

$$\begin{array}{r} ??? \\ 6\overline{)543} \end{array}$$

THINK

$6 \times 100 = 600$
Since 543 is less than 600, the quotient will have two digits.

$$\begin{array}{r} XX \\ 6\overline{)543} \end{array}$$

$$\begin{array}{r} ?X \\ 6\overline{)543} \end{array}$$

To find the leading digit, think of some basic facts that are multiples of 6.

$6 \times 8 = 48$
$6 \times 9 = 54$ 6×9 tens $= 54$ tens or 540

Estimate. $\begin{array}{r} 90 \\ 6\overline{)540} \end{array}$

About 90 trees are needed for each grade.

More Examples

A. $\begin{array}{r} XX \\ 4\overline{)236} \end{array}$ $4 \times 5 = 20$ $\begin{array}{r} 60 \\ 4\overline{)240} \end{array}$
$4 \times 6 = 24$

Since 23 is closer to 24, use 4×6.

B. $\begin{array}{r} XXX \\ 7\overline{)5,124} \end{array}$ $7 \times 7 = 49$ $\begin{array}{r} 700 \\ 7\overline{)4,900} \end{array}$
$7 \times 8 = 56$
$7 \times 6 = 42$

TRY THESE

Find the number of digits in each quotient. Then estimate.

1. $3\overline{)61}$ **2.** $3\overline{)610}$ **3.** $3\overline{)6,100}$ **4.** $4\overline{)79}$ **5.** $4\overline{)790}$

Exercises

Estimate.

1. 3)127   **2.** 2)43   **3.** 8)712   **4.** 6)182   **5.** 8)492

6. 4)133   **7.** 7)291   **8.** 9)625   **9.** 5)716   **10.** 8)583

11. 6)567   **12.** 8)624   **13.** 2)199   **14.** 4)1,752   **15.** 5)1,941

16. 827 ÷ 9   **17.** 4,362 ÷ 6   **18.** 347 ÷ 8   **19.** 347 ÷ 9   **20.** 5,423 ÷ 7

21. 1,374 ÷ 2   **22.** 2,516 ÷ 6   **23.** 8,888 ÷ 9

PROBLEM SOLVING

24. Four buses were driven to a track meet. There were fifty-two students on each bus. Find the total number of students on the four buses.

25. A hardware store has 136 feet of rope. About how many 8-foot ropes can be cut from this length?

26. In 1 day, the Towson Library had 1,234 books checked out and 1,185 books checked in. How many more books were checked out?

27. Kelly is making 3 dresses. She needs 58 yards of fabric for all 3 dresses. About how much fabric will she use for each dress?

MIND BUILDER

Logical Thinking

Find each missing divisor or dividend.

    12 R1   10 R2   25   23 R1   18 R2

1. 2)■   **2.** ■)52   **3.** ■)75   **4.** 2)■   **5.** ■)74

6.9 Dividing Greater Numbers

Objective: to divide multidigit dividends by one-digit numbers

The Millers harvested 4,854 bushels of corn. If the same number of bushels was harvested from 3 fields, what was the number of bushels from each field?

The Millers estimate that it is more than 1,000 but less than 2,000. Explain why.

Step 1	Step 2	Step 3	Step 4
Divide the thousands.	**Divide the hundreds.**	**Divide the tens.**	**Divide the ones.**
$\begin{array}{r} 1 \\ 3\overline{)4{,}854} \\ -3\downarrow \\ \hline 18 \end{array}$	$\begin{array}{r} 1{,}6 \\ 3\overline{)4{,}854} \\ -3 \\ \hline 18 \\ -18 \\ \hline 05 \end{array}$	$\begin{array}{r} 1{,}61 \\ 3\overline{)4{,}854} \\ -3 \\ \hline 18 \\ -18 \\ \hline 05 \\ -3\downarrow \\ \hline 24 \end{array}$	$\begin{array}{r} 1{,}618 \\ 3\overline{)4{,}854} \\ -3 \\ \hline 18 \\ -18 \\ \hline 05 \\ -3 \\ \hline 24 \\ -24 \\ \hline 0 \end{array}$
1 thousand + 8 hundreds = 18 hundreds	0 hundreds + 5 tens = 5 tens	2 tens + 4 ones = 24 ones	

There were 1,618 bushels of corn from each field.

More Examples

A.
$$\begin{array}{r} 532 \text{ R5} \\ 6\overline{)3{,}197} \\ -30\downarrow \\ \hline 19 \\ -18\downarrow \\ \hline 17 \\ -12 \\ \hline 5 \end{array}$$

B.
$$\begin{array}{r} 2{,}291 \\ 2\overline{)4{,}582} \\ -4\downarrow \\ \hline 05 \\ -4\downarrow \\ \hline 18 \\ -18\downarrow \\ \hline 02 \\ -2 \\ \hline 0 \end{array}$$

C.
$$\begin{array}{r} 4{,}137 \text{ R8} \\ 9\overline{)37{,}241} \\ -36\downarrow \\ \hline 12 \\ -9\downarrow \\ \hline 34 \\ -27\downarrow \\ \hline 71 \\ -63 \\ \hline 8 \end{array}$$

TRY THESE

Find the quotient.

1. $8\overline{)608}$ 2. $7\overline{)467}$ 3. $5\overline{)6,485}$ 4. $9\overline{)7,479}$

Exercises

Divide.

1. $2\overline{)128}$ 2. $5\overline{)470}$ 3. $2\overline{)175}$ 4. $7\overline{)665}$ 5. $6\overline{)500}$

6. $9\overline{)584}$ 7. $6\overline{)186}$ 8. $4\overline{)144}$ 9. $2\overline{)137}$ 10. $9\overline{)207}$

11. $6\overline{)3,568}$ 12. $8\overline{)3,297}$ 13. $4\overline{)8,587}$ 14. $5\overline{)3,557}$ 15. $3\overline{)9,368}$

Write *true* or *false*.

★ 16. $400 \div 5 < 600 \div 5$

★ 17. $3,303 \div 3 > 6,963 \div 3$

★ 18. $1,275 \div 3 < 1,275 \div 5$

★ 19. $6,060 \div 3 > 6,060 \div 6$

PROBLEM SOLVING

20. The divisor is 9. What is the greatest remainder possible? What is the least remainder possible?

21. Sharon made 144 cookies for a party. Her brother took 2 dozen of the cookies. How many cookies are left for the party?

22. Howard wants to buy a sundae. He can choose chocolate or vanilla ice cream. He can also choose chocolate, cherry, or pineapple topping. How many different kinds of sundaes can he choose? List them all.

23. A truck carried 2,955 pounds of sand. How many more pounds of sand are needed to fill the truck to a total of 4,200 pounds?

24. Yuri spent $8.75 for 5 roses. How much change does he receive from $10.00?

★ 25. What number multiplied by itself gives the product 1,225?

6.10 Zeros in the Quotient

Objective: to determine when to put a 0 in the quotient

A state park ordered 619 flower plants for 3 different sections of the park. How many flowers will be planted in each section? To find the answer, divide 619 by 3.

THINK   Estimate. $3\overline{)619}$ $\begin{array}{c} XXX \\ 3\overline{)619} \end{array}$ $3 \times 2 = 6$ $\begin{array}{c} 200 \\ 3\overline{)600} \end{array}$

Step 1	Step 2	Step 3
Divide the hundreds.	Divide the tens.	Divide the ones.
$\begin{array}{r} 2 \\ 3\overline{)619} \\ -6\downarrow \\ \hline 01 \end{array}$	$\begin{array}{r} 20 \\ 3\overline{)619} \\ -6\downarrow \\ \hline 01\downarrow \\ -0\downarrow \\ \hline 19 \end{array}$ Why is a 0 put in the quotient?	$\begin{array}{r} 206\ R1 \\ 3\overline{)619} \\ -6\downarrow \\ \hline 01\downarrow \\ -0\downarrow \\ \hline 19 \\ -18 \\ \hline 1 \end{array}$
0 hundreds + 1 ten = 1 ten	1 ten + 9 ones = 19 ones	

There will be 206 flowers planted in each section and 1 flower left over. How does the estimate help you remember to put the 0 in the quotient?

Compare the answer to the estimate. Is the answer reasonable?

> **REMEMBER**
>
> Divide
> Multiply
> Subtract
> Compare
> Bring down

More Examples

A. $\begin{array}{r} 20\ R4 \\ 7\overline{)144} \\ -14\downarrow \\ \hline 04 \\ -\ 0 \\ \hline 4 \end{array}$ Check: $\begin{array}{r} 20 \\ \times\ \ 7 \\ \hline 140 \\ +\ \ \ 4 \\ \hline 144\ \checkmark \text{ It checks.} \end{array}$

B. $\begin{array}{r} 202\ R2 \\ 4\overline{)810} \\ -8\downarrow \\ \hline 01\downarrow \\ -\ 0\downarrow \\ \hline 10 \\ -\ 8 \\ \hline 2 \end{array}$

C. $\begin{array}{r} 1{,}201 \\ 6\overline{)7{,}206} \\ -6\downarrow\downarrow \\ \hline 12\downarrow \\ -12\downarrow \\ \hline 00\downarrow \\ -\ 0\downarrow \\ \hline 06 \\ -\ 6 \\ \hline 0 \end{array}$

TRY THESE

Divide.

1. $6\overline{)27}$ **2.** $2\overline{)616}$ **3.** $6\overline{)244}$ **4.** $4\overline{)404}$ **5.** $5\overline{)754}$

Exercises

Divide.

1. $4\overline{)280}$ **2.** $4\overline{)41}$ **3.** $2\overline{)604}$ **4.** $5\overline{)152}$ **5.** $9\overline{)905}$

6. $3\overline{)927}$ **7.** $2\overline{)211}$ **8.** $8\overline{)85}$ **9.** $7\overline{)76}$ **10.** $3\overline{)121}$

11. $8\overline{)640}$ **12.** $5\overline{)904}$ **13.** $3\overline{)602}$ **14.** $3\overline{)329}$ **15.** $9\overline{)995}$

16. $5,403 \div 5$ **17.** $1,819 \div 9$ **18.** $70,622 \div 6$ **19.** $5,641 \div 7$ **20.** $84,532 \div 6$

Complete.

★ **21.** quotient × divisor + remainder = _____

PROBLEM SOLVING

22. A case of tennis balls has 24 cans. Each can has 3 balls. Ted buys 4 cases of tennis balls for a tournament. How many tennis balls did he buy?

23. The state park planted 412 trees evenly in 4 different sections of the park. How many trees were planted in each section?

24. Bethany has a cherry tree in her backyard. This year it produced 28 pints of cherries. Bethany wants to use as many of the cherries as possible so they will not go to waste. She found a Web site with recipes and ideas of things she could make using the cherries. Using the information she found to the right, what are some combinations of items she can make to use as many of her cherries as possible? Explain your reasoning. (*Hint:* She can make double or triple batches of any recipe, too.)

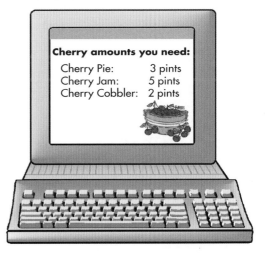

Cherry amounts you need:

Cherry Pie: 3 pints
Cherry Jam: 5 pints
Cherry Cobbler: 2 pints

6.11 Dividing with Money

Objective: to divide money amounts by one-digit numbers

Mrs. Benson pays $27.68 for 4 hanging baskets. Each basket costs the same amount. How much did each basket cost?

THINK An estimate is $28 ÷ 4, or $7.

To find out, divide $27.68 by 4.

Step 1	Step 2
$$\begin{array}{r} \$. \\ 4\overline{)\$27.68} \end{array}$$ Divide dollars and cents like you divide whole numbers. Put the decimal point and dollar sign in the quotient.	Divide. $$\begin{array}{r} \$6.92 \\ 4\overline{)\$27.68} \\ -24 \\ \hline 3\,6 \\ -3\,6 \\ \hline 08 \\ -8 \\ \hline 0 \end{array}$$

Step 1 is very important.

You need to place the decimal point and dollar sign in the quotient. Numbers that show money are different from whole numbers. For example, $2.40 is not equal to 240.

Each hanging basket cost $6.92.

More Examples

A.
$$\begin{array}{r} \$0.75 \\ 4\overline{)\$3.00} \\ -2\,8 \\ \hline 20 \\ -20 \\ \hline 0 \end{array}$$

B.
$$\begin{array}{r} \$1.07 \\ 6\overline{)\$6.42} \\ -6 \\ \hline 0\,4 \\ -\,0 \\ \hline 42 \\ -\,42 \\ \hline 0 \end{array}$$

C.
$$\begin{array}{r} \$0.09 \\ 5\overline{)\$0.45} \\ -45 \\ \hline 0 \end{array}$$

D.
$$\begin{array}{r} \$1.34 \\ 4\overline{)\$5.36} \\ -4 \\ \hline 1\,3 \\ -1\,2 \\ \hline 16 \\ -16 \\ \hline 0 \end{array}$$

TRY THESE

Divide.

1. $2\overline{)\$7.52}$

2. $5\overline{)\$4.00}$

3. $6\overline{)\$6.36}$

4. $3\overline{)\$0.24}$

5. $5\overline{)\$67.65}$

Exercises

Divide.

1. $3\overline{)\$4.71}$ 2. $6\overline{)\$3.00}$ 3. $5\overline{)\$5.25}$ 4. $8\overline{)\$0.56}$

5. $7\overline{)\$9.73}$ 6. $3\overline{)\$11.01}$ 7. $4\overline{)\$8.64}$ 8. $2\overline{)\$2.14}$

9. $2\overline{)\$47.28}$ 10. $5\overline{)\$4.25}$ 11. $9\overline{)\$0.63}$ 12. $7\overline{)\$58.45}$

13. $\$6.85 \div 5$ 14. $\$6.00 \div 6$ 15. $\$7.42 \div 7$ 16. $\$0.72 \div 8$

17. $\$37.17 \div 9$ 18. $\$77.79 \div 3$ 19. $\$29.05 \div 5$ 20. $\$92.20 \div 4$

Write *true* or *false*.

21. In $\$3.92 \div 2$, your quotient must have a dollar sign.

PROBLEM SOLVING

22. Ezra spent $4.56 for 6 plants. Then he spent $3.90 for 6 planters. How much did Ezra spend?

23. How much does each T-shirt cost?

24. Charlie spent $6.30 for 9 plants. He gave the clerk a $20 bill. How much change did Charlie receive?

25. Stuart finds a sale on goldfish. The goldfish are 6 for $2.16. How much is each goldfish?

TEST PREP

26. Anne wanted to take 7 of her friends with her to Gary's Go-Cart Park. Tickets are $12.65 each. How much will it cost her to spend the day with her friends at Gary's Go-Cart Park?

    a. $88.55   **b.** $100.40   **c.** $101.20   **d.** $104.00

6.12 Problem-Solving Strategy: Choosing the Operation

Objective: to choose the correct operation to solve a problem

A tram lift can take 24 people per trip to the top of the mountain. It makes 8 trips up the mountain per day. If the tram is full for every trip, how many people are transported in a day?

After reading a problem, decide if you must add, subtract, multiply, or divide.

Remember what you need to do for each operation:
- ■ *Add* to combine groups or to find a total.
- ■ *Subtract* to take away from a group or to compare groups.
- ■ *Multiply* to combine groups that have the same number.
- ■ *Divide* to separate the total number into equal groups.

| 1. READ | Find the total number of people transported up the mountain in a day. You are given the number of people per trip and the number of trips per day. |

| 2. PLAN | Since you are combining groups that have the same number, multiply. |

| 3. SOLVE | **THINK** An estimate is 25×8, or 200. |

$$\begin{array}{r} 3 \\ 24 \\ \times\ 8 \\ \hline 192 \end{array}$$

If the tram lift is full for every trip, it transports 192 people per day.

| 4. CHECK | Check by changing the order of the factors. |

$$\begin{array}{r} 24 \\ 8)\overline{192} \\ -16 \\ \hline 32 \\ -32 \\ \hline 0 \end{array} \qquad \begin{array}{r} 8 \\ \times\ 24 \\ \hline 32 \\ +\ 160 \\ \hline 192\ \checkmark \end{array}$$

192 people ÷ 8 trips = 24 people per trip

TRY THESE

Choose the operation, and then solve. Explain how you knew what to do.

1. Karen had some books. She gave 33 books to her sister. Now Karen has 17 books. How many books did she have?

2. Ernie has 64 trading cards. He wants to give the same number of cards to his 4 friends. How many trading cards will each friend receive?

Solve

1. At Camp Black Hawk, six boys and a counselor live in each cabin. How many people live in six cabins?

2. Walt spent $27 for 9 books. Each book cost the same amount. How much did each book cost?

3. Charles traveled 7 miles to see a movie. Doug traveled 21 miles. How many more miles did Doug travel?

4. Mirna and her father went fishing. Mirna's father said the round-trip to the lake would be 18 miles. How far was it to the lake?

5. Mrs. Carr collects Beatrix Potter figurines. She has 36 figures on 4 shelves. Each shelf has the same number of figurines. How many figurines are on each shelf?

6. Ana's mother bought 10 tickets to the movies for Ana and her friends. The tickets cost $2.50 each. How much did Ana's mother pay for the tickets?

7. Carl made 12 peanut butter and jelly sandwiches. He is packing 6 lunch bags. Each bag will have the same number of sandwiches. How many sandwiches will he put in each bag?

8. Carl wants to add an equal number of carrot strips to each of the 6 lunch bags. He has 20 carrot strips. How many carrot strips will he put in each bag?

9. After the tram lift ride, Dimas goes to the gift shop. He wants to buy 1 cap and 5 postcards to remember his ride. Using the information to the right, how much will the items cost in all?

10. Yvonne bought 72 buttons for a sewing project. There were 9 buttons in each package. How many packages did she buy?

CHAPTER 6 REVIEW

LANGUAGE and CONCEPTS

Write the correct letter for each part of the division equation.

1. quotient

2. divisor

3. dividend

$$40 \div 5 = 8$$
$$\uparrow \quad \uparrow \quad \uparrow$$
$$\text{A} \quad \text{B} \quad \text{C}$$

Answer each question.

4. What is a fact family?

5. What is a remainder?

6. Which one is a division idea?

   a. Kim and Kevin have 12 model train cars altogether. Kim has 4. How many does Kevin have?

   b. Kevin has 6 model train cars. Kim has 3 times as many train cars. How many does Kim have?

   c. Kevin and Kim have 12 model train cars. They group them with 3 in each group. How many groups are there?

   d. Kim has 10 model train cars. She gave 5 train cars to her brother. How many does she have left?

SKILLS and PROBLEM SOLVING

Write each fact family. (Section 6.1)

7. 9, 4, 36   8. 64, 8, 8   9. 3, 8, 24   10. 6, 5, 30

11. Write the most challenging fact family you can.

Complete the division equation. Write the related multiplication fact. (Sections 6.1–6.2)

12. $39 \div 3 = \blacksquare$ 13. $24 \div 4 = \blacksquare$ 14. $64 \div 8 = \blacksquare$

Estimate. (Section 6.8)

15. $4\overline{)72}$ 16. $2\overline{)65}$ 17. $5\overline{)159}$ 18. $8\overline{)651}$ 19. $6\overline{)375}$

20. $3\overline{)175}$ 21. $6\overline{)\$4.26}$ 22. $3\overline{)928}$ 23. $5\overline{)4,735}$ 24. $7\overline{)6,278}$

Find the quotient. (Sections 6.3–6.4 and 6.6–6.11)

25. $3\overline{)150}$ 26. $6\overline{)180}$ 27. $4\overline{)240}$ 28. $7\overline{)4,900}$ 29. $8\overline{)6,400}$

30. $7\overline{)94}$ 31. $5\overline{)\$95}$ 32. $5\overline{)555}$ 33. $6\overline{)675}$ 34. $3\overline{)805}$

35. $3\overline{)61}$ 36. $2\overline{)\$8.06}$ 37. $6\overline{)485}$ 38. $5\overline{)545}$ 39. $4\overline{)802}$

40. $9\overline{)\$50.04}$ 41. $5\overline{)\$26.15}$ 42. $7\overline{)\$49.49}$ 43. $4\overline{)\$6,592}$ 44. $9\overline{)\$3,681}$

Solve. (Sections 6.5 and 6.12)

45. Daphne had 75 tulip bulbs to plant. She planted 5 bulbs in each row. How many rows did she plant?

46. Jeff needs 15 quarts of potting soil. How many 4-quart bags should he buy?

47. Eileen spent $57 on 3 gifts. Each gift cost the same amount. How much did each gift cost?

48. Mr. Lee needs 30 hot dogs for the cookout. How many packages of 8 hot dogs should he buy?

CHAPTER 6 TEST

Complete the division equation.

1. $56 \div 8 = \blacksquare$ **2.** $40 \div 5 = \blacksquare$ **3.** $21 \div 3 = \blacksquare$

Divide mentally.

4. $2\overline{)80}$ **5.** $3\overline{)606}$ **6.** $4\overline{)320}$ **7.** $5\overline{)1,000}$ **8.** $6\overline{)4,200}$

Divide.

9. $3\overline{)50}$ **10.** $9\overline{)\$99}$ **11.** $7\overline{)\$371}$ **12.** $8\overline{)\$9.04}$ **13.** $5\overline{)884}$

14. $8\overline{)\$0.56}$ **15.** $7\overline{)6,428}$ **16.** $4\overline{)4,327}$ **17.** $2\overline{)\$74.50}$ **18.** $8\overline{)\$27.52}$

Estimate.

19. $6\overline{)492}$ **20.** $3\overline{)237}$ **21.** $5\overline{)6,325}$ **22.** $7\overline{)2,941}$ **23.** $9\overline{)8,036}$

Solve. For each problem, write the operation you used to help you solve the problem.

24. McCormick Elementary had a bake sale. In 1 hour, the cake booth sold 9 cakes for $55.80. Each cake cost the same amount. How much did each cake cost?

25. Carrie needs 3 cups of flour for each cake. How many cakes can she make if she has 17 cups of flour?

26. Donna needs 42 birthday invitations. The invitations are sold in packages of 5. How many packages should Donna buy?

27. Taylor bought a desk for $37.58 and a chair for her room. She spent $55.73. How much was the chair?

Divisibility Rules

A number is **divisible** by another number if it can be divided exactly, with no remainder.

There are shortcuts to help you find out if a number is divisible by another number.

Divisible by...	Shortcut
2	If a number is an even number, it is divisible by 2. Look in the ones place for 0, 2, 4, 6 or 8.
3	Add up all the digits in the number. (Example: 69 → 6 + 9 = 15) Look at the sum. If the sum can be divided evenly by 3, the original number is divisible by 3.
5	If a number ends in a 5 or a 0, it is divisible by 5.
10	If a number ends in 0, it is divisible by 10.

For each exercise, use a different chart of numbers 1–130.

1. Shade all numbers divisible by 2. Can you find a pattern?

2. Shade all numbers divisible by 3. Can you find a pattern?

3. Shade all numbers divisible by 4. Can you find a pattern?

4. Shade all numbers divisible by 5. Can you find a pattern?

5. Shade all numbers divisible by 10. Can you find a pattern?

6. Do you notice any numbers that are divisible by all five preceding numbers?

1	2	3	4	5	6	7	8	9	10
11	12	13	14	15	16	17	18	19	20
21	22	23	24	25	26	27	28	29	30
31	32	33	34	35	36	37	38	39	40
41	42	43	44	45	46	47	48	49	50
51	52	53	54	55	56	57	58	59	60
61	62	63	64	65	66	67	68	69	70
71	72	73	74	75	76	77	78	79	80
81	82	83	84	85	86	87	88	89	90
91	92	93	94	95	96	97	98	99	100
101	102	103	104	105	106	107	108	109	110
111	112	113	114	115	116	117	118	119	120
121	122	123	124	125	126	127	128	129	130

CUMULATIVE TEST

1. Choose the best estimate.

814
× 5

a. 400

b. 4,000

c. 4,500

d. none of the above

2. Which digit is in the hundreds place?

91,217,635

a. 5

b. 6

c. 7

d. 3

3. Name the property.
14 × 13 = 13 × 14

a. Associative Property

b. Commutative Property

c. Zero Property

d. none of the above

4.

68,947
+ 32,684

a. 89,531

b. 90,521

c. 101,631

d. none of the above

5.

$58.65
−   23.98

a. $25.86

b. $35.67

c. $34.67

d. none of the above

6. 6)8,379

a. 943 R3

b. 1,128 R1

c. 1,396 R3

d. none of the above

7. How much money is shown?

a. $2.25

b. $7.75

c. $8.25

d. none of the above

8. What number is a multiple of 8?

a. 31

b. 46

c. 32

d. 59

9. Johnny has $1.50 in his pocket. He has only dimes and nickels. He has the same number of each. How many of each coin does he have?

a. 8 of each coin

b. 10 of each coin

c. 12 of each coin

d. none of the above

10. Bob, Karen, and Shirley help their grandfather pick apples. If they pick the same amount each day, how many apples would they pick together in 3 days?

a. 462

b. 1,124

c. 1,386

d. none of the above

Child	Apples Picked Daily
Bob	156
Karen	162
Shirley	144

Dividing by Two-Digit Numbers

Mikaela Mroczynski
North Carolina

7.1 Division Patterns

Objective: to divide by recalling facts of multiplication

A store needs to order 50 brooms. The supplier sells the brooms in boxes of 10. How many boxes should the store order?

Divide 50 by 10.

THINK   Multiplication and division are related.

$$10 \times \blacksquare = 50$$
$$10 \times 5 = 50$$

$$\begin{array}{r} 5 \\ 10\overline{)50} \\ -50 \\ \hline 0 \end{array}$$

The store needs to order 5 boxes.

More Examples

A.

$$\begin{array}{r} 7 \\ 40\overline{)280} \\ -280 \\ \hline 0 \end{array}$$

$40 \times \blacksquare = 280$

Put the 7 in the ones place.

B.

$$\begin{array}{r} 30 \\ 30\overline{)900} \\ -90 \\ \hline 00 \end{array}$$

$30 \times \blacksquare = 900$

Put 3 in the tens place and 0 in the ones place.

▶ Be sure to put the quotient in the correct place-value position.

TRY THESE

Divide.

1. $20\overline{)80}$ $20 \times \blacksquare = 80$

2. $10\overline{)70}$

3. $30\overline{)210}$

4. $50\overline{)300}$

5. $50\overline{)350}$

6. $60\overline{)360}$

7. $90\overline{)810}$

8. $60\overline{)180}$

Exercises

Divide.

1. $100\overline{)100}$ 2. $40\overline{)160}$ 3. $50\overline{)150}$ 4. $20\overline{)600}$ 5. $60\overline{)4,200}$

6. $20\overline{)40}$ 7. $60\overline{)420}$ 8. $30\overline{)240}$ 9. $10\overline{)90}$ 10. $90\overline{)270}$

11. $70\overline{)1,400}$ 12. $80\overline{)560}$ 13. $60\overline{)480}$ 14. $20\overline{)1,800}$ 15. $30\overline{)2,700}$

16. $20\overline{)1,400}$ 17. $50\overline{)4,000}$ 18. $80\overline{)7,200}$ 19. $40\overline{)3,200}$ 20. $70\overline{)4,200}$

21. $3,000 \div 60$ 22. $4,900 \div 70$ 23. $45,000 \div 90$ 24. $64,000 \div 8$

PROBLEM SOLVING

25. The secretary needs to order 720 pens. The pens are sold in boxes of 80. How many boxes should she order?

26. Arthur wants to order 80 candles. There are 20 candles to a box. How many boxes should he order?

27. Mrs. Redmond has 120 cookies, and there are 30 children at the neighborhood picnic. How many cookies should each child receive?

★ 28. Meg spent $120 on invitations. She bought 60 invitations. How much did each invitation cost?

MIXED REVIEW

29. Christy got a new kitten. She made a table of its growth each week for the first 6 weeks she had it. Use graph paper to make a line graph of the information in the table.

Christy's Kitten's Growth

Week Number	1	2	3	4	5	6
Weight in Ounces	5	8	12	14	16	19

7.2 Estimating Quotients

Objective: to use rounding and estimation to divide mentally; to use compatible numbers to estimate quotients

Mike ordered 779 hamburgers for the cookout. The hamburgers came in 41 packages. About how many hamburgers were in each package?

To estimate, follow these steps:

- Round the **divisor** to its greatest place-value position.
- Round the **dividend** to a multiple of the divisor.
- Then divide.

$$41\overline{)779} \longrightarrow 40\overline{)800}^{\,20} \quad 40 \times \blacksquare = 800$$

> **Compatible numbers** are numbers near the numbers in your original problem that are multiples of the divisor.

Round 41 to 40.
Round 779 to 800. 800 is a compatible number to 40 because 800 is a multiple of 40.

Find $800 \div 40$ by using basic division facts.
Dividing these numbers is an easier, simpler problem.

About 20 hamburgers were in each package.

When you look for a compatible number, you are looking for a number that can easily be divided. Your answer is an estimate.

More Examples

Notice the compatible numbers.

A. $33\overline{)69} \longrightarrow 30\overline{)60}^{\,2}$

$30 \times \blacksquare = 60$

B. $57\overline{)494} \longrightarrow 60\overline{)480}^{\,8}$

$60 \times \blacksquare = 480$

C. $22\overline{)782} \longrightarrow 20\overline{)800}^{\,40}$

$20 \times \blacksquare = 800$

TRY THESE

Estimate. Look for compatible numbers.

1. $51\overline{)98}$　　　**2.** $61\overline{)549}$　　　**3.** $29\overline{)642}$　　　**4.** $28\overline{)680}$

$50 \times \blacksquare = 100$　　$60 \times \blacksquare = 540$

Exercises

Estimate. Look for compatible numbers.

1. $34\overline{)89}$　　　**2.** $64\overline{)132}$　　　**3.** $58\overline{)496}$　　　**4.** $14\overline{)46}$

5. $27\overline{)785}$　　　**6.** $38\overline{)793}$　　　**7.** $76\overline{)651}$　　　**8.** $49\overline{)462}$

9. $34\overline{)65}$　　　**10.** $91\overline{)653}$　　　**11.** $51\overline{)405}$　　　**12.** $36\overline{)267}$

13. $48\overline{)564}$　　　**14.** $52\overline{)362}$　　　**15.** $61\overline{)447}$　　　**16.** $43\overline{)215}$

★**17.** $2,784 \div 8$　　★**18.** $7,209 \div 79$　　★**19.** $9,791 \div 28$　　★**20.** $2,170 \div 70$

21. Estimate the quotient for $5,581 \div 26$.

PROBLEM SOLVING

22. A school bus can hold 58 students. About how many buses would be needed to hold 423 students?

23. Mark has saved 375 bottles for recycling. He put 18 bottles in each box. About how many boxes did he fill?

24. Ann pays $42.62 for a new coat and $13.63 for a hat. How much change does she receive from $100.00?

25. Mr. Smith ordered 42 packages of tulip bulbs. There were 13 bulbs in each package. About how many bulbs did he have to plant?

CONSTRUCTED RESPONSE

26. Mrs. Miles has 2 guinea pigs, 3 dogs, 6 cats, 2 horses, and 1 snake. Each animal eats 1 bag of pet food each week. How many bags of pet food does she need to purchase in a month if there are 4 weeks in a month? Explain your answer.

7.3 Dividing by Multiples of 10

Objective: to divide multi-digit dividends by multiples of 10

There are 258 tickets left for the summer sports festival. Ty wants to send 30 tickets to some of the local gyms. To how many gyms can Ty send tickets? Will there be any tickets left over?

Divide 258 by 30.

Estimate.
$270 \div 30 = 9$

Step 1	Step 2	Step 3
Divide.	Multiply.	Subtract.
$$\begin{array}{r} 8 \\ 30\overline{)258} \end{array}$$	$$\begin{array}{r} 8 \\ 30\overline{)258} \\ 240 \end{array}$$	$$\begin{array}{r} 8\ \text{R18} \\ 30\overline{)258} \\ -240 \\ \hline 18 \end{array}$$
$30 \times$ ■ is equal to 258 or a little less. $30 \times 8 = 240$ ✓ $30 \times 9 = 270$		Is 18 less than 30? yes Do you have anything to bring down? no So, 18 is the remainder.

> If you have a remainder when you divide, the remainder must be less than the divisor.

He can send 8 gyms 30 tickets each. He will have 18 tickets left over.

> Remember your division steps.
> Divide, Multiply, Subtract, Compare, Bring Down

More Examples

A. $$\begin{array}{r} 6\ \text{R7} \\ 40\overline{)247} \\ -240 \\ \hline 7 \end{array}$$

Estimate.
$40 \times$ ■ $= 240$

B. $$\begin{array}{r} 3\ \text{R8} \\ 30\overline{)98} \\ -90 \\ \hline 8 \end{array}$$

Estimate.
$30 \times$ ■ $= 90$

TRY THESE

Divide.

1. $50\overline{)264}$

2. $70\overline{)289}$

3. $40\overline{)93}$

4. $60\overline{)517}$

$50 \times$ ■ is about 250 or less.

Exercises

Divide.

1. $60\overline{)105}$ **2.** $20\overline{)83}$ **3.** $40\overline{)96}$ **4.** $50\overline{)214}$ **5.** $50\overline{)167}$

6. $90\overline{)892}$ **7.** $10\overline{)86}$ **8.** $50\overline{)487}$ **9.** $80\overline{)152}$ **10.** $70\overline{)238}$

11. $10\overline{)94}$ **12.** $70\overline{)647}$ **13.** $90\overline{)300}$ **14.** $20\overline{)68}$ **15.** $40\overline{)186}$

16. $253 \div 30$ **17.** $493 \div 70$ **18.** $632 \div 80$ **19.** $559 \div 80$

Correct the mistake in each problem.

20.
$$\begin{array}{r} 7 \\ 30\overline{)227} \\ -220 \\ \hline 7 \end{array}$$

21.
$$\begin{array}{r} 4 \\ 90\overline{)460} \\ -360 \\ \hline 100 \end{array}$$

PROBLEM SOLVING

22. Sara earns $20 for each project she completes. How many projects must she complete to earn $150?

23. Brenda needs to buy 100 bolts. They are sold in packages of 30. How many packages of bolts does she need to buy?

24. Joe earns $8.00 an hour. How much money does he earn in 40 hours?

25. Find the average of the numbers: 9, 12, 14, 8, 7, 2, 4, 5, 1, and 18.

CONSTRUCTED RESPONSE

26. There are 32 water bottles in a case. Clayton needs to purchase enough water for 130 people to each have a bottle. How many cases will Clayton need to purchase? Explain your reasoning.

27. When you solve a word problem and you get a remainder, describe how the remainder can affect your answer to the problem. Give an example to explain your reasoning.

Problem Solving

Checkers Anyone?

You are given a set of 24 red checkers and a set of 56 black checkers. Arrange the checkers into piles. All piles must have the same number of checkers in them. Each pile may have only red or only black checkers. What is the greatest number of checkers that each pile can have? How many piles of each are there?

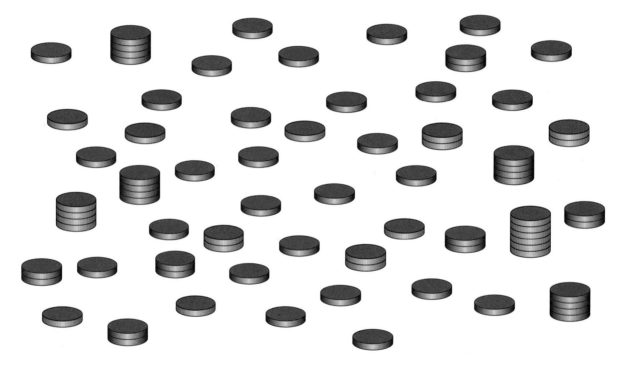

Extension

1. Suppose you are given a set of 12 black checkers and a set of 54 red checkers. What is the greatest number of checkers that each pile can have? How many piles of each are there?

2. Work with other pairs of numbers to solve problems like the preceding one.

Solve.

1. 45×10

2. 561×17

3. $987,000 \times 20$

4. $450 \div 10$

5. $561 \div 30$

6. $987,000 \div 20$

Write in words and in expanded form.

7. $23,456$

8. $1,900,333$

9. $432,987,023$

Solve.

10. Katie has 78 socks. Given they all have a match, how many pairs of socks does Katie have?

11. Davis mows his neighbor's lawn once a week. He gets paid $7.00 each week. How much money will he have after 9 weeks?

12. George collects baseball cards. Each package of cards contains 14 cards and costs 99¢. How much does each card cost?

13. Annie wants to go to a concert. The tickets cost $87.50 each. Annie has $168. Does she have enough money to purchase two tickets to the concert?

Use the following information to complete problems 14–17.

The store received a shipment of T-shirts. There are 16 red-and-white-striped shirts, 34 yellow-and-blue-striped shirts, and 53 white shirts.

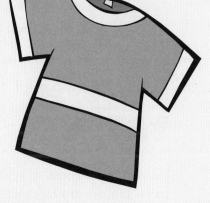

14. Use graph paper to make a bar graph of this information.

15. Are there more striped shirts or solid shirts?

16. The store plans to sell each shirt for $21.95. How much money will the store make by selling all of the T-shirts?

17. The store has a sale on T-shirts. They are 2 for $35.00. Caroline needs to buy 14 shirts. How much money will Caroline spend? Is she saving money by purchasing the T-shirts during the sale? If so, how much?

7.4 Problem-Solving Strategy: Identifying Missing or Extra Facts

Objective: to determine whether a problem can be solved by identifying missing or extra information

4 miles to work
2 miles to the store
6 miles home

Mr. Toyo rides the cable car 4 miles to work, 2 miles to the store, and 6 miles home. He does this 6 times a week. It costs 45¢ to ride. How many miles does he ride on the cable car each day?

Sometimes a problem has more facts than you need to solve it; sometimes it does not have enough facts. You need to examine the question to see what it is asking and whether you have enough information to solve the problem.

1. READ Find the number of miles Mr. Toyo rides on the cable car each day. You know how many miles he rides each time.

2. PLAN Find the total of 4 miles, 2 miles, and 6 miles. You do not need to know the cost of riding the cable car or how many times he rides in a week. Write an equation.

$$4 + 2 + 6 = n$$

3. SOLVE $4 + 2 + 6 = 12$

Mr. Toyo rides on the cable car for 12 miles each day.

4. CHECK Add the numbers in a different order. What property of addition is this?

$$6 + 4 + 2 = 12 ✓$$

TRY THESE

Read the problem. Decide if the story has extra facts or missing facts. List those facts. Do not try to solve.

1. Last month, Mrs. Snyder read some books and watched some movies. This month, she read some books. How many more books did she read this month than last month?

2. Mr. Lowell repaired 4 cable cars. He used 7 cans of red paint, 5 cans of green paint, and 3 cans of oil. How many more cans of red paint than green paint did he use?

Solve

Solve.

1. To get to work, Jan rode a cable car 10 blocks. Then she walked 2 blocks. She traveled the same way home. How far did she travel to work and back?

2. Sarah spent 3 hours sightseeing. She rode a cable car 4 miles. After lunch, she went sightseeing for another 4 hours. How many hours did she spend sightseeing?

Choose the letter of the correct missing fact for each problem. Then solve.

3. Jeff earned $4 for babysitting. He earned extra money for washing dishes. How much money did he earn in all?

4. Gloria bought wood and nails to build a doghouse. The wood cost $11.45. How much did she spend in all?

> Here are the facts!
> a. The nails cost $2.15.
> b. The movie started at 2:30 P.M.
> c. He earned $2.00 for washing dishes.
> d. The hoop cost $16.50.
> e. Wanda's birthday is April 25.

5. Max bought a basketball hoop. He gave the clerk $20.00. How much change did he receive?

6. Joe's birthday is 2 days after Wanda's birthday. What day is Joe's birthday?

7. The movie was 2 hours long. At what time was the movie over?

TEST PREP

Choose the fact needed to solve the problem.

8. Janine pays $7.80 for a blouse and $17.50 for a skirt. How much change should she receive?
   a. the price of a skirt
   b. the price of a blouse
   c. the amount given to the clerk

7.5 Dividing by Two-Digit Numbers

Objective: to divide multi-digit dividends by two-digit numbers

The candy store has 318 pieces of bubble gum. The jars hold 42 pieces each. How many jars can be filled with bubble gum?

Divide 318 by 42.   Estimate. 320 ÷ 40 = 8

Step 1	Step 2
Divide the hundreds and tens. $42\overline{)318}$ Since 42 > 31, the quotient has 0 hundreds and 0 tens.	Divide the ones. $\begin{array}{r} 7\ R24 \\ 42\overline{)3\overset{2\ 11}{\cancel{1}8}} \\ -294 \\ \hline 24 \end{array}$ $\begin{array}{r}\overset{1}{42}\\ \times\ \ 7\\ \hline 294\end{array}$ Since 24 < 42, the quotient is correct.

Seven jars can be filled. There will be 24 pieces left over.

More Examples

A. $\begin{array}{r} 4 \\ 46\overline{)2\overset{1\ 10\ 11}{\cancel{11}}} \\ -184 \\ \hline 27 \end{array}$ Estimate.
200 ÷ 50 = 4
Is 27 < 42? $\begin{array}{r} 4\ R27 \\ 46\overline{)211} \\ -184 \\ \hline 27 \end{array}$

B. $\begin{array}{r} 2\ R8 \\ 44\overline{)96} \\ -88 \\ \hline 8 \end{array}$ Check: $\begin{array}{r} 44 \\ \times\ 2 \\ \hline 88 \end{array}$ $\begin{array}{r} 88 \\ +\ 8 \\ \hline 96\ \checkmark \end{array}$

Check your work.

Step 1	Step 2
Multiply the quotient and divisor. $\begin{array}{r}\overset{2}{46}\\ \times\ 4\\ \hline 184\end{array}$	Add the remainder. $\begin{array}{r}\overset{1\ 1}{184}\\ +\ 27\\ \hline 211\end{array}$ Is 211 the dividend? Yes, your answer is correct.

Gum Balls 5¢
Hard Candy 4¢
Jelly Beans 1¢

TRY THESE

Divide.

1. $17\overline{)74}$ 2. $32\overline{)85}$ 3. $25\overline{)175}$ 4. $45\overline{)158}$

Exercises

Divide.

1. $52 \overline{)108}$ 2. $67 \overline{)83}$ 3. $34 \overline{)96}$ 4. $48 \overline{)407}$ 5. $74 \overline{)559}$

6. $64 \overline{)592}$ 7. $35 \overline{)164}$ 8. $21 \overline{)110}$ 9. $79 \overline{)617}$ 10. $17 \overline{)148}$

11. $130 \div 26$ 12. $97 \div 15$ 13. $143 \div 26$ 14. $420 \div 62$ 15. $172 \div 34$

16. $842 \div 94$ 17. $147 \div 31$ 18. $124 \div 29$

PROBLEM SOLVING

19. Marley has 332 baseball cards. He wants to buy albums that hold 42 cards each. How many albums could he fill?

20. The owner of a toy store needs to put 256 boxes of wooden soldiers on shelves. Each shelf holds 31 boxes. How many shelves does he need to put all the boxes away?

TEST PREP

21. Krisanne runs 84 miles in 14 days. How many miles does she run each day?

   a. 5 miles   **b.** 16 miles   **c.** 7 miles   **d.** 6 miles

MIXED REVIEW

Divide.

22. $5 \overline{)93}$ 23. $6 \overline{)71}$ 24. $4 \overline{)35}$ 25. $3 \overline{)34}$

26. $2 \overline{)46}$ 27. $5 \overline{)903}$ 28. $6 \overline{)731}$ 29. $9 \overline{)913}$

7.6 Two-Digit Quotients

Objective: to divide multi-digit dividends by two-digit
numbers to form two-digit quotients

The fabric store received 960 yards of cotton fabric. Each roll of cloth has
40 yards of fabric on it. How many rolls of fabric did the store receive?

Divide 960 by 40.   Estimate. $800 \div 40 = 20$

Step 1	Step 2	Step 3
Divide the hundreds.	Divide the tens.	Divide the ones.
$40\overline{)960}$	$\begin{array}{r} 2 \\ 40\overline{)960} \\ -80 \\ \hline 16 \end{array}$ $\begin{array}{r} 20 \\ 40\overline{)800} \end{array}$	$\begin{array}{r} 24 \\ 40\overline{)960} \\ -80\downarrow \\ \hline 160 \\ -160 \\ \hline 0 \end{array}$
Since $9 < 40$, the quotient has 0 hundreds.	Since $16 < 40$, the quotient is correct.	Since $0 < 40$, the quotient is correct. Is your answer reasonable?

The store received 24 rolls of fabric.

More Examples

A.
$$\begin{array}{r} 35\text{ R}16 \\ 20\overline{)716} \\ -60\downarrow \\ \hline 116 \\ -100 \\ \hline 16 \end{array}$$

Estimate.
$600 \div 20 = 30$

Check:
$$\begin{array}{r} 1 \\ 35 \\ \times\ 20 \\ \hline 00 \\ +700 \\ \hline 700 \end{array}$$

$$\begin{array}{r} 700 \\ +\ 16 \quad \text{remainder} \\ \hline 716\ \checkmark \end{array}$$

Is 716 your dividend?
Yes, the answer is correct.

B.
$$\begin{array}{r} 30\text{ R}14 \\ 30\overline{)914} \\ -90\downarrow \\ \hline 14 \\ -00 \\ \hline 14 \end{array}$$

Estimate.
$900 \div 30 = 30$

Check:
$$\begin{array}{r} 30 \\ \times\ 30 \\ \hline 00 \\ +900 \\ \hline 900 \end{array}$$

$$\begin{array}{r} 900 \\ +\ 14 \quad \text{remainder} \\ \hline 914\ \checkmark \end{array}$$

TRY THESE

Divide.

1. 30)486 **2.** 50)647 **3.** 20)417 **4.** 30)833 **5.** 70)934

Exercises

Divide.

1. 20)630 **2.** 20)614 **3.** 40)640 **4.** 40)890 **5.** 40)550

6. 60)894 **7.** 70)845 **8.** 30)597 **9.** 90)813 **10.** 10)309

11. 5,312 ÷ 50 **12.** 6,954 ÷ 30 **13.** 7,055 ÷ 10 **14.** 9,150 ÷ 20

PROBLEM SOLVING

15. Last year, 750 children attended karate class. If each class had 20 students, how many classes were held?

16. Jonathan had a total of 930 points on ten spelling tests. What was his average grade on each spelling test?

CONSTRUCTED RESPONSE

17. Is the quotient correct? Explain your reasoning.

793 ÷ 40 = 19 R31

MID-CHAPTER REVIEW

Divide.

1. 40)646 **2.** 81)928 **3.** 70)350 **4.** 51)836 **5.** 26)218

Solve.

6. There are 279 pansies to be planted in 31 pots. Each pot is the same size. How many pansies should be planted in each pot?

7. Forty-three scouts want to sell 309 boxes of cookies. How many boxes should each scout sell?

7.7 Dividing Money

Objective: to divide money amounts by two-digit whole numbers

Fourth grade students sold 37 tickets to their play. They made $296.
How much did each ticket cost?

Divide $296 by 37. Estimate using compatible numbers.

$$300 \div 30 = \$10$$

Step 1	Step 2
Divide the hundreds and tens.	Divide the ones.
$\begin{array}{r} \$ \\ 37\overline{)\$296} \end{array}$	$\begin{array}{r} \$8 \\ 37\overline{)\$296} \\ -296 \\ \hline 0 \end{array}$
Put the dollar sign in the quotient. Since 37 > 29, the quotient has 0 hundreds and 0 tens.	

Each ticket cost $8.

Check:
$$\begin{array}{r} \overset{5}{37} \\ \times \$8 \\ \hline \$296 \checkmark \end{array}$$

More Examples

A.
$$\begin{array}{r} \$0.38 \\ 21\overline{)7.98} \\ -63\downarrow \\ \hline 168 \\ -168 \\ \hline 0 \end{array}$$
Put the decimal point and dollar sign in the quotient.

B.
$$\begin{array}{r} \$0.09 \\ 37\overline{)\$3.33} \\ -3.33 \\ \hline 0 \end{array}$$

TRY THESE

Divide.

1. $58\overline{)\$754}$ 2. $40\overline{)\$9.60}$ 3. $70\overline{)\$490}$ 4. $26\overline{)\$9.88}$

5. $12\overline{)\$1.68}$ 6. $31\overline{)\$9.92}$ 7. $13\overline{)\$91}$ 8. $41\overline{)\$861}$

Exercises

Divide.

1. $27\overline{)\$351}$ 2. $27\overline{)\$162}$ 3. $64\overline{)\$576}$ 4. $61\overline{)\$8.54}$ 5. $37\overline{)\$296}$

6. $12\overline{)\$72}$ 7. $86\overline{)\$688}$ 8. $95\overline{)\$23.75}$ 9. $31\overline{)\$73.16}$ 10. $24\overline{)\$720}$

11. $52\overline{)\$18.20}$ 12. $63\overline{)\$5.67}$ 13. $64\overline{)\$832}$ 14. $18\overline{)\$6.84}$ 15. $18\overline{)\$10.08}$

16. $\$8.64 \div 36$ 17. $\$4.68 \div 13$ 18. $\$9.31 \div 49$ 19. $\$16.15 \div 19$ 20. $\$36.25 \div 29$

21. $\$4.45 \div 89$ 22. $\$11.44 \div 52$ 23. $\$81.18 \div 41$

PROBLEM SOLVING

24. Jason sold 22 tickets to the school play. He collected $27.50. How much did each ticket cost?

25. Joe earned $280 for working 40 hours. How much did he earn for each hour?

26. Martin spent $3.00 on 100 pretzels. How much did each pretzel cost?

★ 27. Nathan is selling thermometers for the choir. He makes $0.75 on each thermometer. Nathan has sold 28 thermometers. How many more does he need to sell in order to make $30.00?

28. Mary bought 3 sponges at $0.89 each. How much change should she receive from a $5 bill?

MIND BUILDER

Targeting Division

Copy and complete each target. Divide each number on the target by the number in the middle.

1.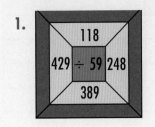

2.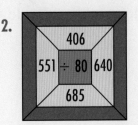

7.8 Problem-Solving Strategy: Looking for a Pattern

Objective: to solve problems by looking for patterns

Two people board the Grand Canyon bus at the first stop. Four people board at the second stop. Six people board at the third stop, and so on. If this pattern continues, how many people will board at the sixth stop?

1. READ Read the problem carefully. You need to know how many people get on the bus at the sixth stop.

2. PLAN Decide how to solve the problem. There is a pattern to the number of people who get on the bus at each stop. Make a table to help find the pattern.

3. SOLVE Solve the problem.

Stop	1	2	3	4	5	6
People	2	4	6	8	10	12

▶ What is the pattern?

4. CHECK Examine, or check, your answer. The pattern continues.

Twelve people will board at the sixth stop.

TRY THESE

Copy and complete the table to see the pattern. Solve.

1. Mona is collecting pennies for the fundraiser. She received 1 penny the first minute, 2 more pennies the second minute, 3 more pennies the third minute, and so on. Complete the table. How much did Mona collect after 9 minutes?

2. Draw an extension of the table. How many minutes will it take Mona to earn over $1.00?

Minutes	Pennies Added	Total Amount Collected
1	1	1¢
2	2	3¢
3	3	6¢
4	4	
5	5	

Exercises

1. Kyle sells lemonade. On the first day, he sold 2 glasses. On the second day, he sold 4 glasses. On the third day, he sold 6 glasses, and so on. For day 5, what was his daily number of glasses? Copy and complete the table.

Day	Daily Number of Glasses	Total Number Sold
1	2	2
2	4	6
3	6	12
4		

2. On which day did Kyle reach a total number sold of 12 glasses?

3. What was Kyle's total number sold at the end of 5 days?

★ 4. What will be Kyle's total number sold at the end of 6 days?

CONSTRUCTED RESPONSE

5. In the preceding lemonade problem, the total number of glasses sold made this series of numbers.

$$2, 6, 12, 20, 30, 42, 56 \ldots$$

What is the next number in the series? What is the pattern? Explain how you discovered the pattern.

MIND BUILDER

Riddle

Divide to answer the riddle. Match each quotient to its letter.

What did the rabbit say when it could not jump the garden fence?

Code

1	2	3	4	5	6	7	8	9
I	H	A	M	R	E	T	O	S

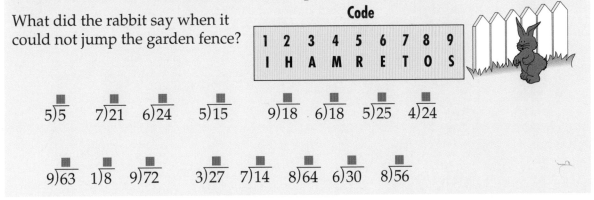

5)5   7)21   6)24   5)15   9)18   6)18   5)25   4)24

9)63   1)8   9)72   3)27   7)14   8)64   6)30   8)56

CHAPTER 7 REVIEW

LANGUAGE and CONCEPTS

Write *true* or *false*. If false, rewrite the sentence to make it correct.

1. 37 is a compatible number for 370.

2. When dividing by multi-digit numbers, the remainder can be more than the divisor.

3. Division can be used to find the number of equal-sized groups or the size of each group.

SKILLS and PROBLEM SOLVING

Divide. (Section 7.1)

4. $20\overline{)480}$ 5. $50\overline{)4,000}$ 6. $90\overline{)7,200}$ 7. $70\overline{)56,000}$

Estimate. Look for compatible numbers. (Section 7.2)

8. $4\overline{)105}$ 9. $8\overline{)675}$ 10. $5\overline{)4,921}$ 11. $7\overline{)2,092}$

Divide. (Sections 7.3 and 7.5–7.7)

12. $30\overline{)280}$ 13. $20\overline{)70}$ 14. $80\overline{)216}$ 15. $70\overline{)507}$

16. $50\overline{)417}$ 17. $90\overline{)\$7.20}$ 18. $46\overline{)148}$ 19. $65\overline{)\$845}$

20. $57\overline{)729}$ 21. $48\overline{)\$768}$ 22. $50\overline{)723}$ 23. $83\overline{)706}$

24. $821 \div 94$ 25. $\$8.16 \div 34$ 26. $\$7.41 \div 19$ 27. $992 \div 97$

Solve. (Section 7.7)

28. Judy bought 2 dozen flower bulbs for $5.28. How much did each bulb cost?

29. Cary Elementary paid $81.25 for 25 students to attend the symphony. How much did the symphony cost per student?

List any missing or extra information. Solve if you can. (Section 7.4)

30. Erin has 13 pairs of pants. She has twice as many shirts as she has dresses. How many shirts does she have?

31. The store received 22 boxes of bathing suits. Each box of bathing suits weighed 15 pounds. Two boxes were damaged. Each box had a different color bathing suit. How many bathing suits were there in all?

32. Emily wants to purchase a dress from her favorite store, as shown to the right. The dress is pink and green. She has $127. Does she have enough money?

Solve. (Section 7.8)

33. In Kasey's room there are two bookcases with five shelves in each bookcase. She has 4 books on the first shelf, 6 books on the second shelf, and 8 books on the third shelf. If this pattern continues, how many books does Kasey have altogether in both bookcases?

Estimate.

1. $6\overline{)659}$ **2.** $90\overline{)927}$ **3.** $50\overline{)670}$ **4.** $53\overline{)368}$ **5.** $20\overline{)140}$

Divide.

6. $26\overline{)\$9.62}$ **7.** $83\overline{)916}$ **8.** $70\overline{)\$840}$ **9.** $26\overline{)628}$ **10.** $79\overline{)614}$

11. $18\overline{)936}$ **12.** $40\overline{)564}$ **13.** $50\overline{)732}$ **14.** $80\overline{)956}$ **15.** $64\overline{)\$7.68}$

16. $32\overline{)979}$ **17.** $56\overline{)\$336}$ **18.** $374 \div 53$ **19.** $780 \div 39$ **20.** $\$420 \div 28$

Solve.

21. A camp director spends $315 on 35 bedsheets for the new bunk beds. How much does each sheet cost?

22. At the grocery store, 2 lb of steak cost $6.80. How much does the steak cost per pound?

Identify any missing or extra information. Solve if possible.

23. In Mary's garden, there are 11 rows of sunflowers. She has 8 rows of daffodils, 7 rows of daisies, and 6 rows of tulips. How many total sunflower plants are in Mary's garden?

24. Sam's backyard has 14 raspberry bushes, some gooseberry bushes, and 3 blackberry bushes. How many berry bushes are there in all in his yard?

25. Juliet's vegetable garden has 7 different kinds of tomato plants. Each plant has 4 stems with 3 tomatoes on each stem. How many tomatoes are growing?

26. Ethel's uncle gave her $20 on Saturday for her birthday. She put half of it in her savings account. She now has $23.50 in her account. How much money did she have in her account before her uncle's birthday gift?

Solve.

27. Kai put 3¢ in his bank on Monday, 6¢ on Tuesday, and 9¢ on Wednesday. If this pattern continues, how much will Kai have in his bank in 2 weeks?

Functions

Jacob has programmed his home computer to multiply numbers by 8. "Multiply by 8" is the rule, or **function**, of his program.

Multiply by 8.

Input	3	5	4	9	8
Output	24	40	32	72	64

Complete each table. Use the rule.

1. Add 125.

Input	61	46	1,265	90	1,345
Output	186	■	■	■	■

2. Divide by 4.

Input	32	64	■	12	■
Output	8	■	7	■	5

3. Add 4. Then subtract 2.

Input	6	29	43	37	72
Output	8	■	■	■	■

4. Subtract 8. Then add 5.

Input	27	13	36	40	84
Output	24	■	■	■	■

Write the rule. Then complete each table.

5.

Input	4	9	6	■	3
Output	28	63	42	70	■

6.

Input	5	7	2	7
Output	20	22	■	■

CUMULATIVE TEST

1. 350 ÷ 16
   a. 22
   b. 21 R14
   c. 23 R16
   d. 16 R22

2. 117 ÷ 3
   a. 38
   b. 41
   c. 39
   d. 93

3. 650 ÷ 10
   a. 65
   b. 10
   c. 6,500
   d. 1

4. Dina needed to buy 14 packages of paper. Each package of paper cost $2.19. How much did Dina spend altogether?
   a. $30.66
   b. $31.67
   c. $30.00
   d. none of the above

5. In the number 876,542, in which place is the 7?
   a. millions
   b. ten thousands
   c. hundred thousands
   d. none of the above

6. Round 56,754 to the nearest thousand.
   a. 56,000
   b. 57,000
   c. 56,700
   d. 56,800

Use the graph to answer questions 7–10.

7. How many class pets are there altogether?
   a. 46
   b. 50
   c. 48
   d. 0

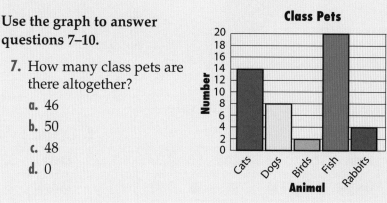

Class Pets

8. If the cats and dogs each ate 2 bags of food per week, how many bags of food would be needed?
   a. 22
   b. 32
   c. 44
   d. none of the above

9. How many more cats and dogs are there than fish?
   a. 2
   b. 3
   c. 4
   d. 0

10. How many more dogs are there than birds and rabbits combined?
    a. 5
    b. 2
    c. 6
    d. 0

CHAPTER 8

Measurement

Ben Daly
Calvert Day School

8.1 Measuring Length with Customary Units

Objective: to compare, estimate, measure, and record linear measurements using customary units

inch (in.)	foot (ft)
	1 ft = 12 in.
yard (yd)	mile (mi)
1 yd = 36 in.	1 mi = 5,280 ft
1 yd = 3 ft	1 mi = 1,760 yd

The **mile** (mi) is used to measure long distances.

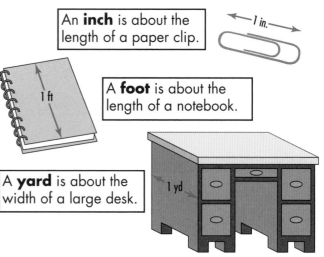

An **inch** is about the length of a paper clip.

A **foot** is about the length of a notebook.

A **yard** is about the width of a large desk.

Examples

Estimate inches.

A.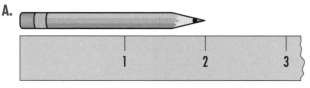

To the *nearest inch,* the pencil is 2 inches (in.) long.

B.

To the *nearest one-half inch,* the pencil is $2\frac{1}{2}$ inches long.

C.

To the *nearest one-fourth inch,* the pencil is $2\frac{1}{4}$ inches long.

More Examples

D. How many inches are in 3 feet?

3 ft = ■ in.

$3 \times 12 = 36$

3 ft = 36 in.

1 ft = 12 in.
You need to change a larger unit to a smaller unit. There will be more units after the change. So, multiply.

E. 2 yd = ■ ft

$2 \times 3 = 6$

2 yd = 6 ft

TRY THESE

Choose the best measurement.

1. length of a "giant step"
2. thickness of this book
3. height of a tall tree
4. distance of a bike race

Measurements
a. 3 feet
b. 30 feet
c. 1 mile
d. 1 inch

Exercises

Estimate and then measure each picture or line.

1.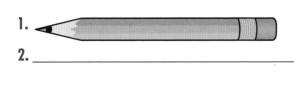
2. _____

Estimate	Nearest inch

3. _____

4. _____

Estimate	Nearest one-half inch

5.

Estimate	Nearest one-fourth inch

6. _____

PROBLEM SOLVING

7. Lance jogged 900 yards. Victor jogged $\frac{1}{2}$ of a mile. Who jogged farther? How much farther?

8. *Estimate* the length of this book. Then measure it to the nearest one-half inch.

8.2 Measuring Capacity with Customary Units

Objective: to compare and estimate liquid measurements using customary units

Cups, **pints**, **quarts**, and **gallons** are used to measure liquids in the customary system of measurement.

Many milk cartons in school are 1-pint size.

2 cups → 1 pint

2 pints → 1 quart

4 quarts → 1 gallon

2 cups (c) = 1 pint (pt)
2 pints = 1 quart (qt)
4 quarts = 1 gallon (gal)

Examples

A. How many cups are in 3 pints?

▇ cups = 3 pints

THINK   2 cups = 1 pint
           2 cups × 3

6 cups = 3 pints

B. How many quarts are in 2 gallons?

▇ quarts = 2 gallons

THINK   4 quarts = 1 gallon
           4 quarts × 2

8 quarts = 2 gallons

TRY THESE

Name the customary unit that is best to measure each of the following. Write *cup*, *pint*, *quart*, or *gallon*.

1. milk for cereal

2. cream

3. paint for a house

4. small jar of jelly

5. gasoline in tank of a car

6. motor oil

7. soup in a bowl

8. can of vegetable juice

9. water in a fish tank

Exercises

Copy and complete. Remember to multiply when changing from larger to smaller units.

1.

qt	1	2	3
c	4	▣	▣

2.

gal	1	2	3
pt	8	▣	▣

3.

gal	1	2	▣
c	16	▣	48

4. 2 pt = ▣ c

5. 2 qt = ▣ pt

6. 5 qt = ▣ pt

7. 5 gal = ▣ pt

8. 5 qt = ▣ c

9. 6 gal = ▣ c

10. 1 qt 1 pt = ▣ pt

11. 1 gal 2 qt = ▣ qt

12. 1 gal 1 pt = ▣ pt

★ **13.** 1 gal 1 qt = ▣ pt

★ **14.** Which is greater: 28 pints or 3 gallons?

PROBLEM SOLVING

▶ 3 teaspoons = 1 tablespoon
16 tablespoons = 1 cup

Use the information in the box to solve problems 15–17.

15. A recipe calls for 1 cup of sugar. How many tablespoons are in 1 cup? How many teaspoons are in 1 cup?

★ **16.** Suppose you have 16 tablespoons of water. How many pints of water would you have?

17. A recipe calls for 1 tablespoon of baking soda. Deborah wants to make a double batch. She only has a teaspoon. How many teaspoons of baking soda will she need?

18. Sophie is having a pancake feast with 20 people. She is planning on each person eating 4 pancakes. A batch of her recipe makes 16 pancakes. How many batches of her recipe will she need to make to feed everyone? Explain your answer.

MIXED REVIEW

Copy and complete.

19.

Feet	Inches
1	12
2	▣
3	▣
4	▣
5	▣

20.

Yards	Feet
1	3
2	▣
3	▣
4	▣
53	★▣

21.

Inches	Yards
36	1
▣	2
▣	3
▣	4
★▣	25

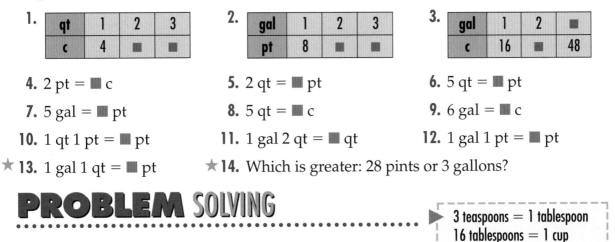

8.3 Measuring Weight with Customary Units

Objective: to compare and estimate weight measurements using customary units

The **ounce** (oz) and **pound** (lb) are used to measure weight. *Very* heavy objects are weighed in **tons** (T).

A loaf of bread weighs about 1 pound.

| 16 ounces (oz) = 1 pound (lb) |
| 2,000 pounds (lb) = 1 ton (T) |

A small car weighs about 1 ton.

Examples

A. 2 T = ■ lb

$2 \times 2,000 = 4,000$

2 T = 4,000 lb

THINK
Larger units changing to smaller units, so multiply.

B. 64 oz = ■ lb

$64 \div 16 = 4$

64 oz = 4 lb

THINK
Smaller units changing to larger units, so divide.

TRY THESE

Name the best customary unit to use to measure each weight.

1. a bag of birdseed

2. a box of cereal

3. a truck

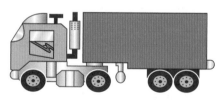

Exercises

Complete each sentence. Use oz, lb, or T.

1. A hamburger weighs about 4 ■.

2. A small car weighs about 1 ■.

3. A hammer weighs about 6 ■.

4. A cat weighs about 7 ■.

5. An elephant weighs about 2 ■.

6. A steak weighs about 8 ■.

Copy and complete.

7. 3 lb = ▦ oz

8. 3 T = ▦ lb

9. 5 lb = ▦ oz

10. 12 T = ▦ lb

11. 18 lb = ▦ oz

★ **12.** 5,000 lb = ▦ T

★ **13.** 32 oz = ▦ lb

★ **14.** 8,000 lb = ▦ T

★ **15.** 56 oz = ▦ lb

PROBLEM SOLVING

16. Tom had 1 pound of ground beef. He used 9 ounces to make hamburgers. How many ounces are left?

17. Mr. Dennison's truck weighs 2 tons. How many pounds does it weigh?

18. A party sandwich weighed 3 pounds. It was 3 feet long. The people at the party ate 44 ounces of the sandwich. How many ounces are left? *Hint*: 3 lb = ▦ oz

19. If an elephant weighs 3,976 pounds, how much more does the 4,153-pound whale at Oceanland Park weigh?

MIXED REVIEW

Measure each line to the nearest one-fourth inch.

20. _____

21. _____

22. _____

Choose the more reasonable measurement.

23. length of a toilet paper roll
a. 10 yd   **b.** 5 in.

24. carton of yogurt   **a.** 1 c   **b.** 1 gal

25. length of an egg carton
a. 12 in.   **b.** 20 ft

26. kitten   **a.** 20 oz   **b.** 20 lb

Complete.

27. 1 yd = ▦ in.

28. 1 c = ▦ pt

29. 5,280 ft = ▦ mi

30. ▦ qt = 2 gal

Solve.

31. Mary wants to knit a scarf that is 2 yd 1 ft long. Each ball of yarn results in 1 ft of knitting. How many balls of yarn does she need?

8.4 Problem-Solving Strategy: Creating a Simpler Problem

Objective: to solve problems by first solving a related, simpler problem

At the grocery store, apples cost $1.25 per half pound. James buys 16 pounds of apples to make apple butter. How much do the apples cost?

 1. READ
Read the problem. You need to find out how much the apples cost.

2. PLAN
Solve a related, simpler problem.
How much would be the total price of the apples if a half pound of apples cost $1.00 and James buys 5 pounds?

THINK A half pound cost $1.00, so a whole pound costs $1.00 + $1.00 = $2.00. Multiply the cost per pound by the number of pounds James buys to get the total cost: $2.00 x 5 = $10.00.

3. SOLVE
Solve the original problem.
If one half pound of apples costs $1.25, one whole pound costs $1.25 + $1.25 = $2.50. Multiply the cost per pound by the number of pounds James buys to get the total cost: $2.50 x 16 = $40.00.

4. CHECK
Estimate.
Round the cost of a half pound of apples to the nearest dollar. A half pound costs about $1.00, so a pound costs about $2.00. Rounding 16 pounds to the greatest place value is 20. $2.00 per pound times 20 pounds is $40.00, so the answer makes sense.

TRY THESE

1. How many multiples of 4 are there between 1 and 100?

2. Luke's father gives him $2.35 every Tuesday and Thursday. His grandmother gives him $3.25 every Monday and Wednesday. Every other Friday he spends $2.19 on a milkshake. How much money does Luke have after 4 full weeks?

Solve

1. Each hour, the elephant ate 37 peanuts and threw 43 peanuts out of his cage. How many peanuts did the trainer give him during the 10 hours the circus was open on Saturday?

2. The circus sold 670 tickets for its final performance. If each ticket cost $11, how much did the circus make?

3. The circus travels all over the country. It traveled 673 miles in June, 756 miles in July, and 1,429 miles in August. How many miles did the circus travel that summer?

★ 4. The traveling circus performs 2 shows each day for 4 days in a row in every city it visits. It takes 3 days to take down the tents and corral the animals. It takes 3 days to travel to the next city. Then it takes 3 days to unload and set up everything for the next 4-day show. How many shows will the circus perform from May 1 to September 30?

MID-CHAPTER REVIEW

Match the equal units of measure.

1. 2 cups	a. 1 mile
2. 64 cups	b. 2 yards
3. 8 cups	c. 4 cups
4. 2 pints	d. 1 ton
5. 6 feet	e. 4 gallons
6. 8 quarts	f. 8 pounds
7. 128 ounces	g. 1 pint
8. 2,000 pounds	h. 4 pints
9. 5,280 feet	i. 2 gallons

8.5 Measuring Length with Metric Units

Objective: to compare, estimate, measure, and record linear measurements using metric units

millimeter (mm)	centimeter (cm)
10 mm = 1 cm	1 cm = 10 mm
decimeter (dm)	**kilometer (km)**
1 dm = 10 cm	1 km = 1,000 m
meter (m)	
1 m = 100 cm; 1 m = 10 dm	

1 cm

A centimeter (cm) is about the width of a fingernail.

A meter (m) is about the width of a door.

1 m

Kilometers are used to measure greater distances. A kilometer (km) is about the distance you can walk quickly in 10 minutes.

Examples

A. How many centimeters are in 2 meters?

2 m = ■ cm
2 × 100 = 200
2 m = 200 cm

THINK
There are 100 cm in 1m.

B. How many kilometers are in 5,000 meters?

5,000 m = ■ km

5,000 ÷ 1,000 = 5

5,000 m = 5 km

1,000 m = 1 km
You need to change a smaller unit to a larger unit. There will be fewer units after the change. So, divide.

TRY THESE

Name the metric unit that is best to measure each of the following.
Write *millimeter, centimeter, meter,* or *kilometer.*

1. distance of a trip

2. length of a screwdriver

3. height of a building

4. length of your thumb

5. distance of a hike

6. thickness of a sandwich

7. height of a flagpole

8. length of a caterpillar

9. width of a grain of rice

Exercises

Estimate. Then measure the length to the nearest centimeter.

1.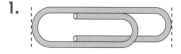

2.

3.

Estimate. Then measure each of the following to the nearest meter.

4. height of a door

5. width of your bed

6. height of a kitchen table

7. length of a window

Copy and complete.

8. 7 m = ■ cm

9. 4 m = ■ cm

10. 20 mm = ■ cm

11. 6 m = ■ cm

12. 66 km = ■ m

13. 800 mm = ■ cm

14. 2 km = ■ m

15. 2,000 m = ■ km

16. 8 km = ■ m

★**17.** 200 cm = ■ m

★**18.** 400 mm = ■ m

★**19.** 7,000 m = ■ km

PROBLEM SOLVING

20. Mr. Karnes ran a 10-kilometer race. How many meters did he run?

21. Terry swam 4 meters underwater. How many centimeters is this?

22. Tracy has a jump rope that is 2 meters long. How many centimeters long is it?

23. Millie ran a 50-meter race. How many centimeters is this?

TEST PREP

Choose the most reasonable measurement.

24. distance jogged **a.** 4 cm **b.** 4 m **c.** 4 km

25. length of a fishing pole **a.** 2 cm **b.** 2 m **c.** 2 km

26. width of a shovel **a.** 30 cm **b.** 30 m **c.** 30 km

27. height of a sesame seed **a.** 1 cm **b.** 1 m **c.** 1 mm

MIND BUILDER

Precise Measurement

A **millimeter** (mm) is a metric unit used to measure shorter or narrower objects or to make a more precise measurement.

This ruler is marked in mm.
1 cm = 10 mm

Measure each line segment. Use a ruler marked in millimeters.

1. _____

2. _____

3. _____

4. _____

8.6 Measuring Capacity with Metric Units

Objective: to compare and estimate liquid measurements using metric units

The pitcher holds about 1 **liter** (L) of lemonade. In the metric system, the liter is used to measure liquid. The liter is the metric unit of capacity.

An eyedropper will hold about 1 **milliliter** (mL) of liquid.

$$1,000 \text{ milliliters (mL)} = 1 \text{ liter (L)}$$

Examples

A. 2 L = ■ mL

    2 × 1,000 = 2,000

    2 L = 2,000 mL

THINK
When changing from a larger unit to a smaller unit, you need to multiply.

B. 53,000 mL = ■ L

    53,000 ÷ 1,000 = 53

    53,000 mL = 53 L

THINK
When changing from a smaller unit to a larger unit, you need to divide.

TRY THESE

Complete.

1. 4 L = ■ mL

2. 23 L = ■ mL

★ **3.** 16,000 mL = ■ L

Choose the more reasonable measurement.

4.

2 L or 2 mL

5.

8 L or 8 mL

6.

4 L or 4 mL

Exercises

Choose the more reasonable measurement.

1.

4 L or 4 mL

2.

355 L or 355 mL

3.

500 L or 500 mL

4. milk in a glass: 30 mL or 30 L

5. pitcher of lemonade: 1 L or 1 mL

6. gas tank in a car: 20 L or 20 mL

7. water in a bathtub: 80 L or 80 mL

8. Which is greater: 4,000 mL or 2 L?

Copy and complete.

9. 3 L = ■ mL

★**10.** 4,000 mL = ■ L

11. 25 L = ■ mL

12. 8 L = ■ mL

13. 10 L = ■ mL

★**14.** 17,000 mL = ■ L

PROBLEM SOLVING

15. Laura measures the water in a bucket. It holds 6,000 mL. How many liters does it hold?

16. Patrick has 26 signed baseballs from players in the major leagues. He can fit 4 on a shelf. How many shelves will he need?

17. Katie collects dolls. She has 3 shelves with 18 dolls on each shelf. How many dolls does she have?

18. A cow drinks about 56 pints of water a day. How many gallons of water does a cow drink?

MIXED REVIEW

Divide.

19. 5)25

20. 6)934

21. 7)53

22. 8)$43.92

Complete.

23. 2 ft = ■ in.

24. 3 yd = ■ ft

25. 1 mi = ■ ft

26. The average of 6, 4, 3, and 3 is ____.

8.7 Measuring Mass with Metric Units

Objective: to compare and estimate mass measurements using metric units

gram (g)	kilogram (kg)
	1 kg = 1,000 g

A dollar bill has a mass of about 1 gram.

A pair of tennis shoes has a mass of about 1 kilogram.

Examples

A. 3 kg = ■ g

3 × 1,000 = 3,000

3 kg = 3,000 g

THINK
When changing a larger unit to a smaller unit, you multiply.

B. 6,000 g = ■ kg

6,000 ÷ 1,000 = 6

6,000 g = 6 kg

THINK
When changing a smaller unit to a larger unit, you divide.

TRY THESE

Name the metric unit that is better to measure each of the following. Write *gram* or *kilogram*.

1. paper clip
2. bowling ball
3. dog
4. 2 dollar bills
5. person
6. chair

Choose the more reasonable mass.

7. a box of crackers
   500 g or 500 kg
8. a computer
   10 g or 10 kg
9. a loaf of bread
   500 g or 5 kg

Exercises

Copy and complete.

1. 2 kg = ■ g
2. 6 kg = ■ g
3. 25 kg = ■ g
4. 30 kg = ■ g
★ 5. 8,000 g = ■ kg
★ 6. 18,000 g = ■ kg

Name the metric unit that is better to measure each of the following. Write *gram* or *kilogram*.

7.

8.

9.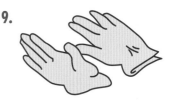

10.

11.

12.

Choose the more reasonable mass.

13. a box of cereal: 500 g or 500 kg

14. a dog: 10 g or 10 kg

15. a bag of oranges: 5 g or 5 kg

16. a pair of socks: 40 g or 4 kg

17. Which is greater: 250 g or 3 kg?

18. Which is greater: 7 kg or 8,250 g?

PROBLEM SOLVING

Use the chart to solve problems 19–21.

19. Miguel has some coins that have a value of 25¢. The coins weigh about 9 grams. What are the coins?

20. Chan has some coins that weigh 12 grams. What is the greatest value of the coins?

★21. Don gets an allowance that weighs 52 grams. What is the least and the most his allowance could be?

Coin	Value	Weight About
Penny	1¢	3 g
Nickel	5¢	5 g
Dime	10¢	2 g
Quarter	25¢	6 g

★ 22. Myra has 16 marbles. One half of them are red. How many marbles are red?

CONSTRUCTED RESPONSE

23. One nickel weighs about 5 g. One dime weighs about 2 g. Kasen believes that on a scale, he can balance 4 nickels with 8 dimes. Is he correct? Why or why not?

Problem Solving

Paint the Cube Red

A cube that measures 3 inches in height, 3 inches in length, and 3 inches in width is painted red on all sides. If the cube is then cut into 1-inch cubes, how many cubes have 3 sides painted red? To help solve the problem, draw diagrams and number the cubes.

Extension

1. Try to find the number of cubes that have 0, 1, 2, 4, 5, and 6 sides painted red.

2. Project
Use sugar cubes or other cube-shaped items. Stack the cubes 4 cubes in height, 4 cubes in length, and 4 cubes in width. Mark each side showing with an **X**. Predict the number of cubes that have 0, 1, 2, 3, 4, 5, and 6 sides marked. Check your prediction by unstacking the cubes and counting.

CUMULATIVE REVIEW

Write in expanded form.

1. 37   **2.** 852   **3.** 1,083   **4.** 46,549

Add or subtract.

5. 5,698
  + 279

6. 24,428
  − 12,977

7. 23,064
  + 56,835

8. 3,273
  − 1,186

9. 945
    836
  + 3,328

Multiply or divide.

10. 3×5 **11.** $8\overline{)32}$ **12.** $8\overline{)48}$ **13.** 9×9

14. 53×2 **15.** 200×9 **16.** $4\overline{)73}$ **17.** 8×624

18. $7\overline{)231}$ **19.** $9 \times 2 \times 4$ **20.** 26×13 **21.** $6\overline{)579}$

22. 271×35 **23.** $4 \times \$3.17$ **24.** $2\overline{)5,437}$ **25.** 40×8

Use the double-bar graph to answer questions 26–27.

26. Whose garden has more flowers: Hope's or Jennifer's?

27. Does Hope have more roses, or does Jennifer have more geraniums and tulips combined?

Solve.

28. A nursery planted 15 rows of trees with 25 trees in each row. How many trees were planted?

29. Mr. Tyler has 426 flowers. If he puts 3 flowers in each vase, how many vases will he use?

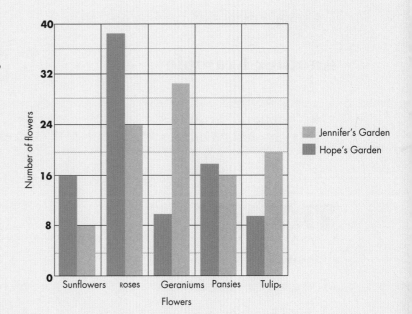

Objective: to add like units of measurement

Jane and Matt want to measure the height of their puppet stage.

They know these measurements:
- ■ floor to the stage—3 feet 4 inches
- ■ stage to the valance—1 foot 4 inches
- ■ valance height—10 inches

```
   3 feet   4 inches     Add like units.
   1 foot   4 inches
+           10 inches
   4 feet  18 inches
```

> How many inches are in a foot? Since 18 inches is more than the number of inches in a foot, rename the inches to feet and inches.

THINK   18 in. = 12 in. + 6 in. or 1 ft 6 in.

4 feet 18 inches ⟶ **4 feet + 1 feet 6 inches** ⟶ **5 feet 6 inches**

The puppet stage is 5 feet 6 inches tall.

Another Example

> **REMEMBER**
> There are 16 ounces in a pound.

```
    4 pounds  13 ounces
+   7 pounds  10 ounces
   11 pounds  23 ounces = 12 pounds  7 ounces
```

THINK   23 oz = 16 oz + 7 oz = 1 lb 7 oz

TRY THESE

Rename.

1. 6 gal 5 qt = 7 gal ■ qt

2. 3 ft 15 in. = 4 ft ■ in.

3. 5 yd 5 ft = 6 yd ■ ft

4. 1 lb 20 oz = 2 lb ■ oz

5. ■ L = 3 L 2,000 mL

★ **6.** ■ qt ■ c = 2 qt 5 c

Exercises

Copy and complete.

1. 1 km = ▦ m

2. 100 cm = ▦ m

3. 1 ft = ▦ in.

4. 1 yd = ▦ ft

5. 1 kg = ▦ g

6. 1 lb = ▦ oz

7. 1 L = ▦ mL

8. 2 c = ▦ pt

9. 1 qt = ▦ pt

Find the sum. Rename, if needed.

10.
```
    5 m 60 cm
+   7 m 60 cm
```

11.
```
    25 km 400 m
+   14 km 750 m
```

12.
```
    2 gal 3 qt
+         3 qt
```

13.
```
    1 qt 1 c
+   2 qt 3 c
```

14.
```
    5 yd 2 ft
+   2 yd 2 ft
```

15.
```
    10 lb 9 oz
+    8 lb 7 oz
```

16.
```
    7 pt 1 c
+   2 pt 1 c
```

17.
```
    8 kg 299 g
+   4 kg 780 g
```

18.
```
    14 gal 2 qt
+    3 gal 3 qt
```

PROBLEM SOLVING

19. Hilary throws a shot put 3 yards 2 feet. She throws it again 4 yards 2 feet. What is the total distance she throws the shot put?

20. Jane and Matt carry 3 puppets in a suitcase. The puppets weigh 12 oz, 15 oz, and 13 oz. The empty suitcase weighs 5 lb. When the puppets are in the suitcase, how much does it weigh?

★ **21.** Maggie walked 1,200 yards. Julia walked 1 mile. Who walked farther? How much farther?

★ **22.** Jack drives 208 km 532 m on the first day of his trip. On the second day, he drives 250 km 641 m. On the third day, he drives only 123 km 132 m. How far does he drive in 3 days?

8.9 Temperature

Objective: to read positive and negative changes in temperature using Celsius and Fahrenheit scales

Claire and Andy plan to paint a fence. Claire reads the paint can.
It states the temperature should be above 50°F in order to paint. Andy
checks the outdoor thermometer.

Andy reports the temperature is 47°F. Can they paint the fence? Why or why not?

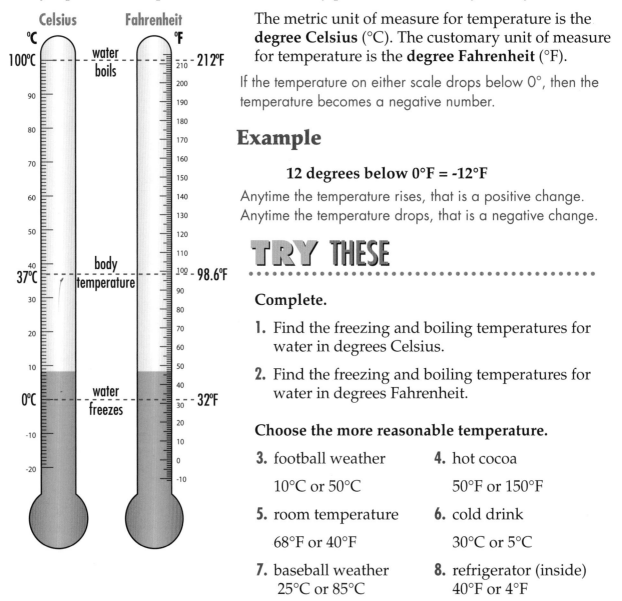

The metric unit of measure for temperature is the **degree Celsius** (°C). The customary unit of measure for temperature is the **degree Fahrenheit** (°F).

If the temperature on either scale drops below 0°, then the temperature becomes a negative number.

Example

12 degrees below 0°F = -12°F

Anytime the temperature rises, that is a positive change.
Anytime the temperature drops, that is a negative change.

TRY THESE

Complete.

1. Find the freezing and boiling temperatures for water in degrees Celsius.

2. Find the freezing and boiling temperatures for water in degrees Fahrenheit.

Choose the more reasonable temperature.

3. football weather

   10°C or 50°C

4. hot cocoa

   50°F or 150°F

5. room temperature

   68°F or 40°F

6. cold drink

   30°C or 5°C

7. baseball weather
   25°C or 85°C

8. refrigerator (inside)
   40°F or 4°F

Exercises

Choose the more reasonable temperature.

1.

   25°F or 65°F

2.

   86°C or 20°C

3.

   0°C or 32°Cz

4. It is 40°F outside. Is that above or below freezing?

5. The temperature of the surface of a lake is 27°F. Can you go swimming?

6. The temperature was 29°C. It rose 5 degrees. What is the temperature now? Is that a positive or negative change?

7. The temperature fell 5°F. It is now 68°F. What was the original temperature? Was that a positive or a negative change?

8. The lowest temperature ever recorded in Alaska was -80°F. This was recorded in January, 1971, at Prospect Creek Camp. Draw a thermometer to show this temperature.

9. Death Valley National Park in California is one of the hottest places on Earth. It had the second-highest temperature ever recorded, 134°F in 1913. Draw a thermometer to show this temperature.

TEST PREP

10. A pipe 10 meters long needs to be cut into pieces 1 meter long. How many cuts need to be made?

   a. 10 cuts
   b. 9 cuts
   c. 11 cuts
   d. There is not enough information.

11. The temperature was 10°F on Monday. On Tuesday it was -8°F. How many degrees did the temperature drop?

   a. 10°F
   b. 15°F
   c. 18°F
   d. none of the above

Objective: to tell time and the passage of time using digital and analog clocks

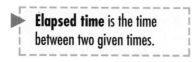

There are two types of clocks.
A **digital clock** shows the time in number form.
An **analog clock** shows the time by using hands on a dial.

> ▶ **Elapsed time** is the time between two given times.

> 60 seconds = 1 minute
> 60 minutes = 1 hour
> 24 hours = 1 day

On Saturday, Tony begins building his model ship at 10:45 A.M. He stops working at 11:05 A.M. How many minutes has he worked on his model?

Tony worked on his model ship for 20 minutes.

Look at the minute hand. Count the minutes from 10:45 to 11:05. From 10:45 A.M. to 11:05 A.M. is 20 minutes.

More Examples

A. What time is 4 hours after 9:00 A.M.?

The time changes to P.M. when the hour hand passes 12 noon.

1:00 P.M.

B. What time is 30 minutes before 8:50 A.M.?

Start at 8:50. Count back 30 minutes.

8:20 A.M.

TRY THESE

Copy and complete.

1.

_____ after _____

2.
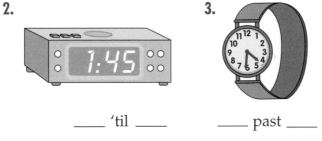

_____ 'til _____

3.

_____ past _____

Exercises

What time is it?

1. 20 minutes after 2:00 P.M.

2. 30 minutes after 3:30 P.M.

3. 20 minutes before 8:00 A.M.

4. 15 minutes before 9:30 P.M.

5. 25 minutes before 6:10 A.M.

6. 15 minutes before 8:55 P.M.

7. 25 minutes after 5:15 A.M.

8. 30 minutes before 11:45 P.M.

What is the elapsed time?

9. 6:00 A.M. and 6:25 A.M.

10. 8:15 P.M. and 9:15 P.M.

11. 8:10 A.M. and 8:50 A.M.

12. 4:25 P.M. and 5:10 P.M.

PROBLEM SOLVING

13. Write the time that is 6 minutes before 8.

14. Draw a clock face that shows 25 minutes after 6:17.

15. How many seconds are in 1 minute and 10 seconds?

16. How many seconds are in 3 minutes?

17. A school day is 6 hours. How many seconds are in 1 school day?

18. How many seconds are in 1 day?

19. Mrs. Thurman put her pie in the oven for 1 hour and 10 minutes. How many minutes will the pie bake? How many seconds?

20. Mark begins to work on his model. He can work for 50 minutes until lunch is ready. At what time will he be called for lunch?

21. Lucy began studying at 3:10 P.M. She studied for 2 hours and 35 minutes. What time was it when she finished?

CHAPTER 8 REVIEW

LANGUAGE and CONCEPTS

Write the letter of the correct word or number to complete each sentence.

1. Inch is to the customary system as centimeter is to the _____ system.

2. Gram is to mass as milliliter is to _____.

3. _____ is a measure of how heavy an object is.

4. A ton is equal to _____ pounds.

5. 212°F is to boiling water as _____°F is to freezing water.

a. capacity
b. 2,000
c. 32
d. weight
e. metric

SKILLS and PROBLEM SOLVING

Choose the more reasonable measurement.
(Sections 8.1–8.3, 8.5–8.7, and 8.9)

6. warm rolls
   a. 30°C   b. 80°C

7. juice in a glass
   a. 250 mL   b. 250 L

8. weight of a telephone
   a. 2 g   b. 2 kg

9. length of a skateboard
   a. 2 in.   b. 2 ft

Copy and complete. (Sections 8.1–8.3 and 8.5–8.7)

10. 3 ft = ■ in.

11. 5 lb = ■ oz

12. 4 pt = ■ c

13. 6 qt = ■ gal ■ qt

14. 2,650 mL = ■ L ■ mL

Find the sum. Rename, if needed. (Section 8.8)

15. $\quad$ 45 cm
     $+\ 2$ m 85 cm

16. $\quad$ 25 gal 1 qt
     $+\ \ 5$ gal 3 qt

17. $\quad$ 7 kg
     $+\ 4$ kg 450 g

Solve. (Sections 8.4 and 8.9)

18. Each actor has 6 costume changes in the first half of the production and 14 changes in the second half. How many costume changes do 12 actors have?

19. Frank and his mother are baking cookies. At which temperature would they bake the cookies: 350°F or 350°C? Explain why.

20. At what temperatures Fahrenheit and Celsius does water freeze?

21. It was 95°F. There was a negative change in temperature of 17°F . Draw a thermometer to show the new temperature.

22. If today is Monday, what day of the week will it be in 50 days?

Write each time. Use numbers and A.M. or P.M. (Section 8.10)

23.

lunchtime

24.

asleep

25.

early evening

What is the elapsed time? (Section 8.10)

26. 3:10 P.M. and 4:05 P.M.

27. 2:35 A.M. and 3:20 A.M.

Solve. (Section 8.10)

28. Rob began his homework at 6:55 P.M. He finished at 8:23 P.M. For how long did Rob work on his homework?

29. The movie starts at 2:15 P.M. and ends at 4:35 P.M. How long is the movie?

30. Cole plays basketball for 1 hour and 20 minutes every Saturday. If practice starts at 11:40 A.M., at what time does Cole finish?

Copy and complete.

1. 14 kg = ■ g
2. 9 ft = ■ in.
3. 8 yd = ■ ft
4. 12 yd = ■ in.
5. 12 m = ■ cm
6. 7 pt = ■ c
7. 8 qt = ■ pt
8. 16 gal = ■ qt
9. 2 gal = ■ pt
10. 3 qt = ■ c
11. 11 T = ■ lb
12. 25 lb = ■ oz
13. 23 m = ■ cm
14. 20 kg = ■ g
15. 15 L = ■ mL
16. 6 yd = 18 ■
17. 8 km = 8,000 ■
18. 4 pt = 8 ■

What is the elapsed time?

19. 3:00 A.M. and 3:45 A.M.
20. 9:05 P.M. and 9:30 P.M.

Solve.

21. Mr. Bradley leaves work at 5:15 P.M. He must drive for 1 hour and 25 minutes. What time does he arrive home?

22. What is the boiling point of water in degrees Fahrenheit? in degrees Celsius?

23. The high temperature for the day was 83°F. The daily low temperature was 69°F. Find the negative change in temperature.

24. How many multiples of 6 are there between 1 and 100?

25. Twenty small square tables will be used for a party. Each table can only seat one person on each side. If the 20 tables are pushed together to make one long table, how many people can sit at the table?

CHANGE OF PACE

Historic Units of Measurement

Long ago people used body parts to measure length. Here are some of the more common units of measure they used.

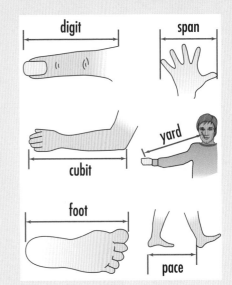

- digit—the length of a finger

- span—the length of a hand from the little finger to the thumb with fingers spread

- cubit—the distance from the elbow to the fingertips

- yard—the distance from the nose to the fingertips with arm outstretched at the side

- foot—the length of a foot

- pace—a double step (Roman)

Measure the following distances using the units given. Make a table of your findings. Compare your measurements with the measurements of a friend.

Object	Unit	My Measurement	Friend's Measurement	Difference
1. Width of a computer screen	Span			
2. Width of a doorway	Foot			
3. Length of a car	Pace			
4. Length of a loaf of bread	Digit			

Complete the following problems.

5. What problems, if any, do you see in using these units of measure in your everyday life?

6. **Research**
Find out more about these and other units of measure used in the past or today. Make a poster to report your research.

1. Complete the fact family.

$7 + 4 = 11$
$11 - 7 = 4$
$11 - 4 = 7$

a. $4 + 7 = 11$
b. $7 - 4 = 3$
c. $11 + 4 = 15$
d. none of the above

2. Which symbol can replace ● to make this true? 45¢ ● $45

a. $<$
b. $=$
c. $>$
d. none of the above

3.
$$\begin{array}{r} \$1.38 \\ \times \quad 7 \\ \hline \end{array}$$

a. $9.16
b. $9.66
c. $42.66
d. none of the above

4. What is the time?

a. 7:00 A.M.
b. 7:20 A.M.
c. 7:40 P.M.
d. none of the above

breakfast

5. What is the elapsed time between 8:45 A.M. and 9:25 A.M.?

a. 20 minutes
b. 30 minutes
c. 40 minutes
d. none of the above

6. About what length is the line segment?

_____

a. 1 cm
b. 2 cm
c. 3 cm
d. none of the above

7. Nina had 70 stickers. She gave away 11 stickers and bought 16 new stickers. How many stickers does she have now?

a. 75 stickers
b. 84 stickers
c. 175 stickers
d. none of the above

Use the stem-and-leaf plot to answer questions 8–9.

Daily Temperatures in May at Noon	
Stem	**Leaf**
6	6 6 7 8 8 9 9
7	2 4 5 5 6 6 7 8 8 9 9
8	1 2 4 4 4 4 5 5 6 8 8
9	0 1

Key: 6|7= 67°F

8. What is the median of daily May temperatures at noon?

a. 76°F
b. 78°F
c. 84°F
d. none of the above

9. What is the mode of daily May temperatures at noon?

a. 76°F
b. 78°F
c. 84°F
d. none of the above

CHAPTER 9

Geometry

Lisa Binner
Missouri

Objective: to identify and name points, lines, line segments, and planes

A **point** represents an exact location in space. A dot is a model of a point.

M
•

Write: M

Say: point M

A **line** is a never-ending, straight path that extends in both directions.

R   S

Write: $\overleftrightarrow{RS}$ or $\overleftrightarrow{SR}$

Say: line RS or line SR

A **line segment** is part of a line. A line segment consists of two endpoints and all the points of the line that are between these points.

A •————————• B

Write: $\overline{AB}$ or $\overline{BA}$

Say: line segment AB or line segment BA

A **plane** is a never-ending, flat surface. A basketball court is a model of a part of a plane. Any three points in a plane (in any order) can be used to name the plane.

Say: plane QPR, plane RPQ, plane PRQ, etc.

TRY THESE

Name two other models for the following.

1. point: *tip of a needle*   a. _____   b. _____

2. line segment: *edge of a book*   a. _____   b. _____

3. part of a plane: *surface of a tennis court*   a. _____   b. _____

Exercises

Draw each of the following.

1. a line through points P and Q

2. a line segment with endpoints B and C

3. point K

4. line AN

5. line segment XY

6. $\overline{XY}$
       7. $\overleftrightarrow{WX}$
       8. $\overline{UV}$
       9. $\overleftrightarrow{CD}$

★**10.** a point on line DA
       ★**11.** a line through L, M, and N

Write all the names of each figure.

12. $\xleftrightarrow{\quad\overset{S}{\bullet}\qquad\overset{C}{\bullet}\quad}$
   13. $\overset{F}{\bullet}\rule{6cm}{0.4pt}\overset{G}{\bullet}$
   14. $\overset{E}{\bullet}$

PROBLEM SOLVING

15. How many line segments can you draw using 4 points?

16. Mr. and Mrs. Peck began building a shed at 8:30 A.M. They finished $6\frac{1}{2}$ hours later. At what time did they finish?

MIXED REVIEW

Write the numbers.

17. multiples of 6: 6, 12, ▪, ▪, ▪, ▪, ▪, ▪

18. multiples of 9: 9, ▪, ▪, 36, ▪, ▪, ▪, ▪

19. common multiple of 2 and 4

20. common multiple of 8 and 9

21. common multiple of 6 and 8

22. common multiple of 6 and 10

9.2 Intersecting, Perpendicular, and Parallel Lines

Objective: to identify intersecting lines; to identify and name perpendicular and parallel lines

Lines that cross are called **intersecting lines**. Line QR intersects line ST at point P.

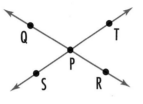

Two intersecting lines that form square corners are called **perpendicular lines**.

Write: $\overleftrightarrow{AB} \perp \overleftrightarrow{CD}$

Say: Line AB is perpendicular to line CD.

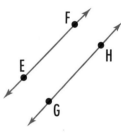

Lines in the same plane that never cross are called **parallel lines**. Parallel lines are the same distance apart at any given point.

Write: $\overleftrightarrow{EF} \parallel \overleftrightarrow{GH}$

Say: Line EF is parallel to line GH.

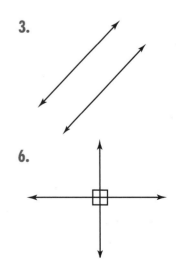

TRY THESE

Describe each pair of lines.

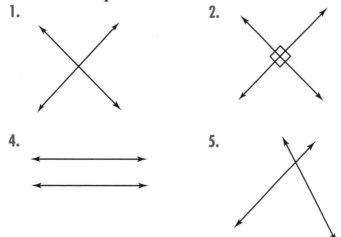

1.

2.

3.

4.

5.

6.

Exercises

Identify the lines as *parallel, intersecting,* or *perpendicular.*

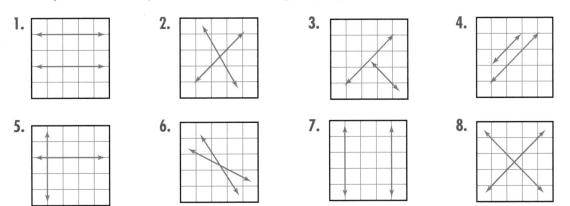

1.

2.

3.

4.

5.

6.

7.

8.

Make three different drawings to show these line relationships. Use graph paper.

9. intersecting   **10.** parallel   **11.** perpendicular

Use the diagram to complete the following problems.

12. Name the line that is parallel to $\overleftrightarrow{MN}$.

13. Name a line that is perpendicular to $\overleftrightarrow{RQ}$.

14. Name a line that intersects $\overline{MP}$ at point O.

15. Name three lines that intersect at point N.

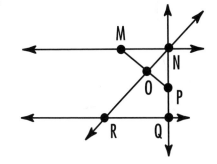

PROBLEM SOLVING

16. What relationship do the sidelines of a football field have?

17. Which of the following letters shown have perpendicular line segments? Which ones have parallel line segments?

A E H K L N

18. Christina spent $23.99 on shoes during a tax-free weekend. How much does she have left over from the $50.00 she brought with her for shopping?

9.3 Rays and Angles

Objective: to identify and name rays and angles; to classify angles

A part of a line that has only one endpoint is called a **ray**. The arrow shows that the ray goes on and on in one direction. When you name a ray, name the endpoint first.

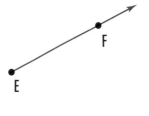

Write: $\overrightarrow{EF}$

Say: ray EF

Two rays that have a common endpoint form an **angle**. The endpoint that they share is called the **vertex**. Angles can be named using the vertex alone or with three letters. (The vertex is always the middle letter.)

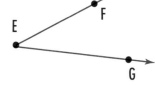

Write: ∠E, ∠FEG, ∠GEF

Say: angle E, angle FEG, or angle GEF

The blades of the scissors open and close as the cuts are made. The sides of the blades form different-sized angles.

In angle XYZ, the vertex is at point Y. The rays YX and YZ are the sides of the angle.

All other angles are compared to right angles.

Angles less than right angles are **acute angles**.

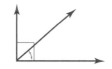

Square corners form special angles called **right angles**. Use the corner of your paper to determine if an angle is a right angle.

Angles greater than right angles are **obtuse angles**.

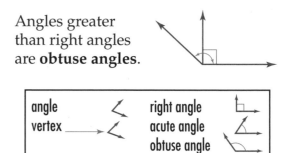

angle	⟨	right angle	⌐
vertex ——→	⟨	acute angle	◿
		obtuse angle	

TRY THESE

1. Name objects in the room that contain right angles, acute angles, and obtuse angles.

Draw each of the following.

2. angle MNO   3. ray AB   4. $\overrightarrow{RQ}$ 5. ∠STU

Compare each angle with a corner of your paper. Identify the angle as *right, acute,* or *obtuse*.

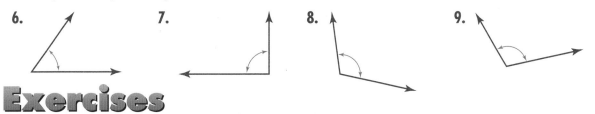

6.   7.   8.   9.

Exercises

Complete the following. Use the figure at the right.

1. The vertex is point ■.

2. One side is ray BD. The other side is ray ■.

3. Name the angle as follows:

   a. one letter, ∠ ■

   b. three letters, ∠ ■ ■ ■ or ∠ ■ ■ ■

Draw each of the following.

4. a right angle   5. an acute angle   6. an obtuse angle

Decide if the angles of these shapes are *acute, obtuse,* or *right*.

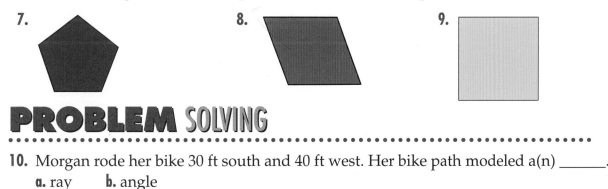

7.   8.   9.

PROBLEM SOLVING

10. Morgan rode her bike 30 ft south and 40 ft west. Her bike path modeled a(n) _____.

    a. ray   **b.** angle

9.4 Polygons

Objective: to identify polygons based on the number of sides

A **polygon** is a closed plane figure with straight sides that do not cross.

Study the names for the following polygons. How many sides does each polygon have?

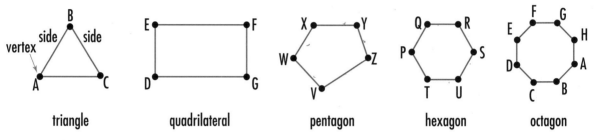

| triangle | quadrilateral | pentagon | hexagon | octagon |

The point where two sides meet is called a **vertex**. Name the vertices (plural of *vertex*) of the pentagon.

TRY THESE
....

Identify each polygon. Write *triangle, quadrilateral, hexagon, octagon,* **or** *pentagon.*

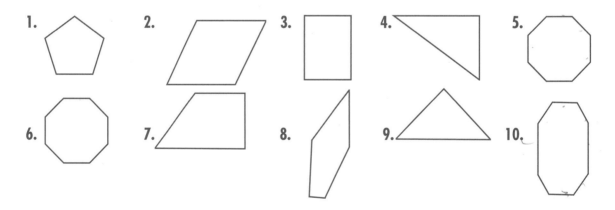

Exercises

Name each polygon.

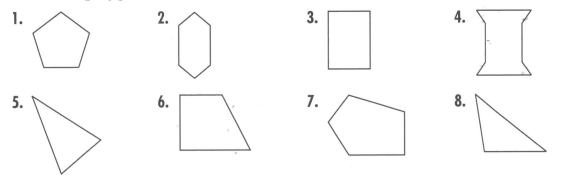

1.

2.

3.

4.

5.

6.

7.

8.

How many sides, angles, and vertices does each polygon have?

9.

10.

11.

12.

13.

PROBLEM SOLVING

Use the figures to the right to solve problems 14–18.

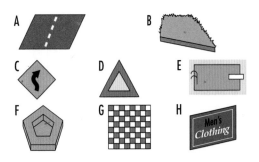

14. Which figures have shapes that are quadrilaterals?

15. Which figures contain right angles?

16. Which figures contain only acute angles?

17. Which figures contain only obtuse angles?

18. Which figures contain both acute and obtuse angles?

19. A hexagon has 6 sides. How many vertices does a hexagon have?

20. An octagon has 8 vertices. How many sides does an octagon have?

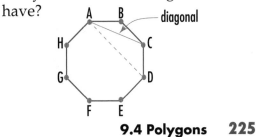

★ 21. A **diagonal** connects two vertices that are not adjacent (not side by side). How many diagonals can be drawn from the vertices of the polygon on the right?

9.5 Quadrilaterals

Objective: to define and classify quadrilaterals

Some quadrilaterals have special names.

The three figures at the right are **parallelograms**. The opposite sides of a parallelogram are parallel.

The red figure and the green figure are **rectangles**. A rectangle is a parallelogram that has four right angles.

The green figure is a **square**. A square is a rectangle that has four congruent sides.

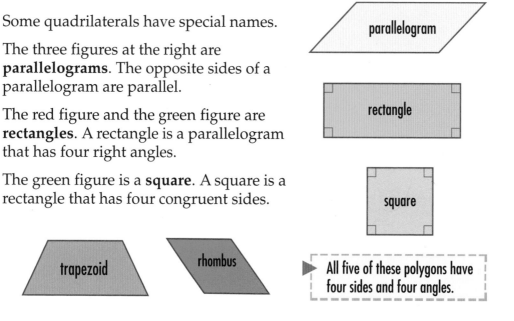

▶ All five of these polygons have four sides and four angles.

There are special ways that you can tell which polygon is which:

- A parallelogram has 2 sets of parallel sides and 2 sets of equal sides.
- A **rhombus** is a parallelogram with 4 equal sides.
- A square and a rectangle are both parallelograms with 4 right angles.
- A square is a rectangle with 4 equal sides.
- A square is a rhombus with 4 right angles.
- A **trapezoid** has only 1 set of parallel sides.

TRY THESE

Name the quadrilateral that each object suggests.

1. 2. 3. 4. 5.

Using the word bank, choose the correct shape for each description. Some problems will have more than one answer.

1. a shape with only one pair of parallel sides
2. a shape with two pairs of equal sides
3. a shape with two pairs of parallel sides
4. a shape with four sides
5. a shape with four equal sides
6. a shape with all right angles

Word Bank	
square	rectangle
trapezoid	parallelogram
rhombus	quadrilateral

PROBLEM SOLVING

7. The highway runs north and south. Sam is driving north, and Pam is driving south. What kind of lines does the highway represent?

8. Draw an obtuse angle.

9. Draw a rectangle. Name one thing that is an example of a rectangle.

10. Name three objects that are shaped like a square.

CONSTRUCTED RESPONSE

11. Sid has a long piece of wire. He cut off two 8-inch pieces and two 4-inch pieces of wire.

   a. How much wire did he have originally? Explain how you know.

   b. What shapes can you make with these four pieces of wire? Draw and label them.

TEST PREP

12. How many factors does 5 have?
   a. 1   b. 3   c. 2   d. 5

13. *True* or *False*: The number 10 has two digits.

9.6 Congruent and Similar Figures

Objective: to distinguish between congruent and similar figures

Maria's grandmother is making a colorful quilt for her bed. To the right is a labeled drawing of the pattern she will be using.

Notice that the quilt will be made from triangles and rectangles of different sizes. Some of the shapes are the exact same size but are turned in different directions. Two or more figures that have the same shape and size are **congruent**. Some examples of congruent triangles and rectangles are shown here:

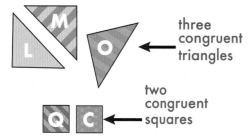

three congruent triangles

two congruent squares

Two or more figures that have the same shape but might differ in size are **similar**.

two similar triangles

TRY THESE

Use the preceding labeled quilt drawing. Write the letters that answer each question.

1. Find two congruent rectangles.
2. Find two similar triangles.
3. Find two congruent triangles with stripes.
4. Find two similar squares.

Exercises

Write the letter of the figure that is congruent to the first one.

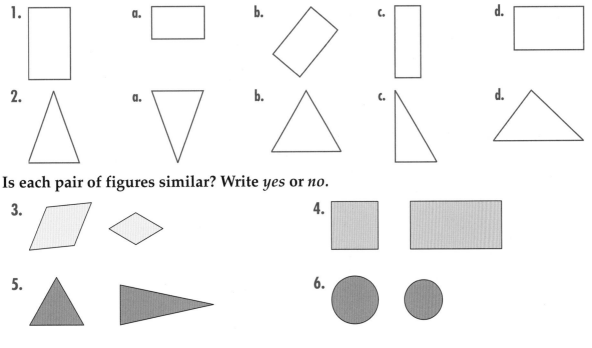

Is each pair of figures similar? Write *yes* or *no*.

PROBLEM SOLVING

7. The sides of figure **A** are congruent to the sides of figure **B**. Are the two figures congruent? Why or why not?

★ **8.** Hal needed to get two color copies of a 4×6-inch photograph made. He wanted to get one that was congruent and one that was similar to the original. How can he get both of these?

TEST PREP

9. Which figure is similar to the blue part of the given figure?

a. b. c.

9.7 Problem-Solving Strategy: Logical Reasoning

Objective: to solve problems using logical reasoning

Which figure does not belong?

Study the four figures. Look for
things that are the same and things that are different.

 1. READ   Find the item that does not belong in this group. All the
items are shapes.

2. PLAN
- The figures are all the same color.
- The figures are all about the same size.
- The figures are all different shapes.

3. SOLVE
- Three figures are polygons.
- One figure is not a polygon.

   The circle does not belong.

4. CHECK   Three shapes have straight lines.
The circle does not have straight lines.

TRY THESE

Give a reason why each shape does not belong.

1. shape A
2. shape B
3. shape C
4. shape D

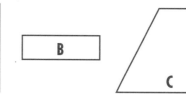

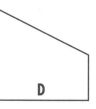

Solve..

Which figure does not belong? Give a reason.

1.
   a.   b.   c.   d.

2.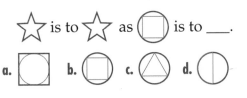
   a.   b.   c.   d.

3.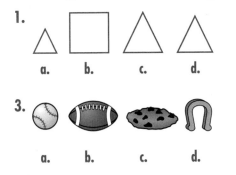
   a.   b.   c.   d.

4.
   a.   b.   c.   d.

Solve.

5. Think how the first two figures are alike. Then pick the figure that is like the third figure.

   is to as is to ___.

   a.   b.   c.   d.

★ 6. How many squares can you find in the following figure?

7. Kathryn hikes 562 yards. How many feet has she hiked?

CONSTRUCTED RESPONSE
....

8. Explain the difference between a circle and a polygon.

MIXED REVIEW
....

Use the following polygons to solve.

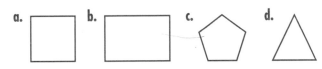
a.   b.   c.   d.

9. Which of the polygons have right angles? acute angles? obtuse angles?

10. Which of the polygons have parallel sides?

11. Name each polygon.

12. Draw a figure similar to each of the polygons.

Objective: to determine whether figures are symmetrical; to find lines of symmetry

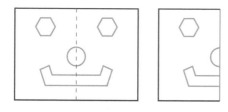

Maria and her father are painting a pattern across the top of the walls with this stencil.

You can see that the stencil can be folded so that both sides match exactly. The pattern in this stencil has a **line of symmetry**.

More Examples

A. This figure has two lines of symmetry.

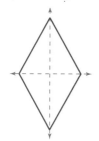

B. This is NOT a line of symmetry.

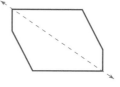

The parts of the figure on either side of the line of symmetry are congruent.

TRY THESE

Is each dashed line a line of symmetry? Write *yes* or *no*.

1. 2. 3. 4. 5.

How many lines of symmetry does each figure have?

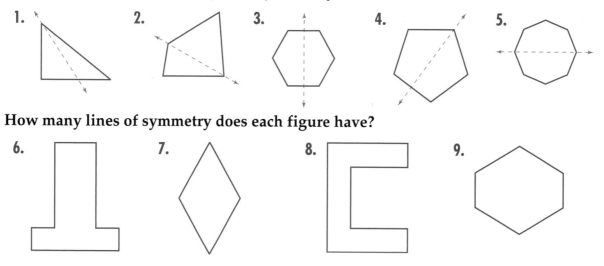

6. 7. 8. 9.

Exercises

How many lines of symmetry does each figure have?

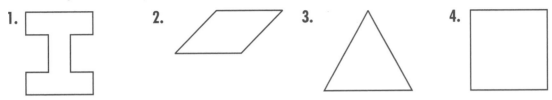

1. 2. 3. 4.

Use graph paper to copy the part of the figure given and the line of symmetry. Complete each figure.

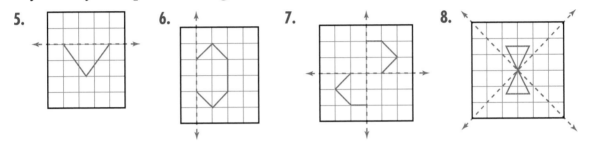

5. 6. 7. 8.

PROBLEM SOLVING

9. List five objects in your room that have at least one line of symmetry.

★ 10. Trace a large circle on a piece of paper. Then cut it out. Fold the circle to see whether there is a line of symmetry. Fold it again and again at different places. How many lines of symmetry do you think a circle has?

MIND BUILDER

Symmetry

Each letter of the alphabet has 0, 1, or 2 lines of symmetry, except the letter O. Trace the alphabet. Then draw the lines of symmetry for each letter.

A B C D E F G H I J K L M N

P Q R S T U V W X Y Z

9.9 Transformations: Slides, Flips, and Turns

Objective: to identify the results of translations, reflections, and rotations of geometric figures

Patterns in material and wallpaper are created by repeating and moving a figure in different ways. The move does not change the size or shape of the figure. So, each repeat forms a figure congruent to the other figures.

A figure can be moved in three ways.

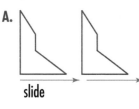

A. slide

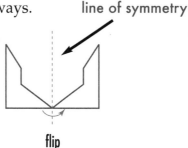

B. line of symmetry

flip

C. turn

Another name for a **slide** is a **translation**.

Another name for a **flip** is a **reflection**.

Another name for a **turn** is a **rotation**.

> A **transformation** in geometry occurs when you move a figure. Flips, slides, and turns are all transformations.

TRY THESE

How were these shapes moved? Write *slide, flip,* or *turn.*

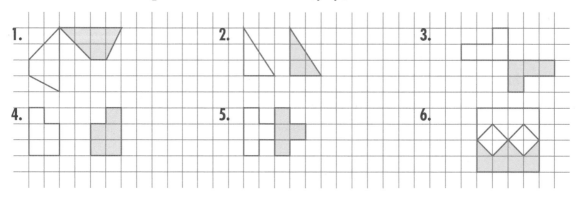

1.

2.

3.

4.

5.

6.

Exercises

1. Design a shape to cut out of cardboard. Place it on graph paper and trace around it. Slide, flip, or turn the shape and trace around it again. Repeat the shape many times to make a pattern.

Move this figure as directed on graph paper. Always start with it in the original position as shown. Draw the figure after the movement.

2. flip

3. turn 90 degrees to the right (a $\frac{1}{4}$ or quarter turn)

4. slide

5. turn 180 degrees to the right (a $\frac{1}{2}$ or two-quarter turn)

PROBLEM SOLVING

6. Name an everyday example of each transformation: flip, slide, and turn.

7. When you put a key in a keyhole and unlock the door, which transformation names that movement?

8. When a fan is on, what transformation is made by the blades?

CONSTRUCTED RESPONSE

9. Which transformation is a mirror image: a slide, a flip, or a turn? Explain.

TEST PREP

10. Which of these does NOT describe a square?
    a. a quadrilateral
    b. three-dimensional
    c. a rectangle
    d. a figure with four 90-degree angles
    e. a polygon

Problem Solving

Billy Bob's Boxes

Billy Bob needs a pattern for an open-top box. The pattern must be in one piece. He can only fold on the dotted lines. Predict which of the patterns Billy Bob can use and how many open-top box patterns he can choose from.

Copy the patterns on large graph paper, cut them out, and fold them to test whether you can make an open-top box.

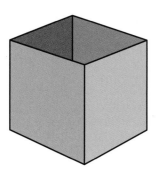

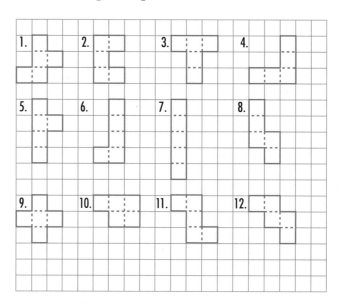

Extension

1. Which patterns will not form an open-top box? Explain why each of these patterns will not work.

2. These patterns are called **pentominoes**. Explain what you think the word means. Look up the definition in the glossary. Does your definition match the definition in the glossary?

CUMULATIVE REVIEW

Compute.

1. $87.52
    − 49.69

2. 6)573

3. 2,937
    + 5,208

4. 6,754
    ×   7

5. 3)38

6. 14 × 6 × 7

7. $7.15 × 24

8. 327 + 739 + 675

Estimate.

9. 317 × 4

10. 541 × 27

11. 72 × 56

Identify the lines as *parallel,* *perpendicular,* **or** *intersecting.*

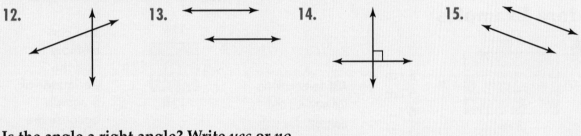

12.

13.

14.

15.

Is the angle a right angle? Write *yes* **or** *no.*

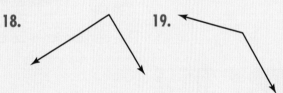

16.

17.

18.

19.

Write the ordered pair for each point.

20. B

21. E

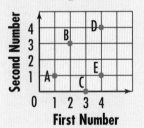

Solve.

22. Mike earns $1.85 an hour doing yard work. One month he worked 27 hours. How much did he earn?

9.10 Perimeter

Objective: to find the perimeter of polygons

Grady wants a fence around a yard in his model train village. He put a string around the yard to measure it. The string measured 20 centimeters, so Grady needs 20 centimeters of fence.

The distance around the outside of a polygon is the **perimeter**. To find the perimeter, add the lengths of the sides.

Meter means to measure; *peri* means around.

More Examples

A.

8 cm
5 cm   5 cm
8 cm

8 cm
5 cm
8 cm
+ 5 cm
26 cm

▶ Add measurements the way you add numbers. The units *must* be the same.

The perimeter is 26 cm.

B.

12 ft
12 ft   12 ft
12 ft

12 ft
× 4 sides
48 ft

You can multiply to find the perimeter of a *square* because the sides are equal.

The perimeter is 48 ft.

TRY THESE

Find the perimeter.

1. 8 cm   16 cm   8 cm
    16 cm

    16 cm
    8 cm
    16 cm
    + 8 cm

2. 6 ft   6 ft   6 ft
    6 ft

    6 ft
    × 4 sides

3. 7 m   21 m
    16 m   10 m

4. 27 mi
    4 mi   6 mi   10 mi
    5 mi

5. 7 in.
    7 in.   7 in.
    7 in.

Exercises

Find the perimeter.

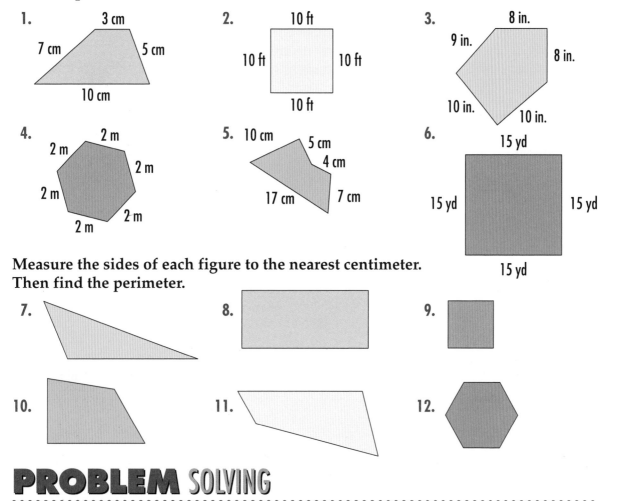

1. 3 cm, 7 cm, 5 cm, 10 cm

2. 10 ft, 10 ft, 10 ft, 10 ft

3. 8 in., 9 in., 8 in., 10 in., 10 in.

4. 2 m, 2 m, 2 m, 2 m, 2 m, 2 m, 2 m

5. 10 cm, 5 cm, 4 cm, 17 cm, 7 cm

6. 15 yd, 15 yd, 15 yd, 15 yd

Measure the sides of each figure to the nearest centimeter. Then find the perimeter.

7.

8.

9.

10.

11.

12.

PROBLEM SOLVING

13. A square has a perimeter of 8 inches. How long is each side?

14. The perimeter of a rectangle is 38 cm. The length of one side is 6 cm. What are the lengths of the other sides?

15. The perimeter is 20 inches. Each side is 4 inches long. How many sides are there?

16. Draw as many rectangles as you can that have a perimeter of 24 cm. How many can you draw? (Use only whole numbers.)

★**17.** A polygon has eight sides. Two of the sides are 6 cm. Two of the sides are 5 cm. The other sides are 4 cm, 3 cm, 8 cm, and 10 cm. What is the perimeter and the name of the polygon?

9.11 Area

Objective: to find the area of two-dimensional shapes

You want to cover the floor of the clubhouse with tiles. To cover the floor, you need to know its **area**. The area of a region is the number of square units needed to *cover* the region.

Each tile on the clubhouse floor measures 1 foot on each side. The area of each tile is 1 square foot. Count the squares to find the area.

The area of the clubhouse floor is 55 square feet.
Why do you include the word *square*?

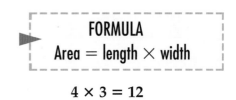

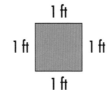

Another Example

What is the area of this rectangle?

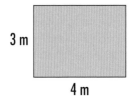

The area of this rectangle is 12 square meters.

You can find the area of a square or a rectangle by multiplying the length times the width.

▶ ┌─────────────────────────────┐
　 │　　　　**FORMULA**　　　　 │
　 │ Area = length × width │
　 └─────────────────────────────┘

$$4 \times 3 = 12$$

TRY THESE
• •

Find the area of each figure by counting.

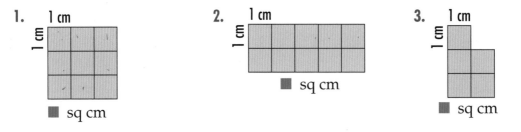

1. 　▉ sq cm

2. 　▉ sq cm

3. 　▉ sq cm

Exercises

Find the area. Write the formula.

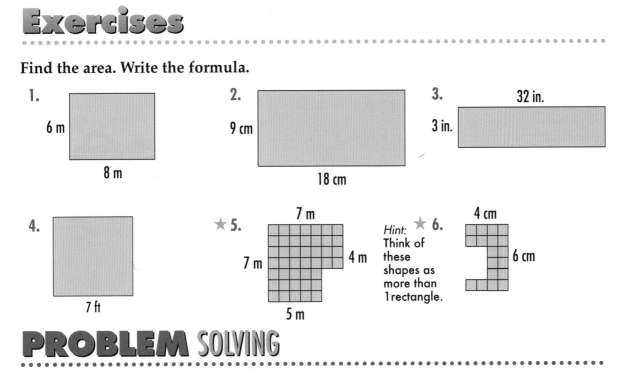

1. 6 m, 8 m

2. 9 cm, 18 cm

3. 32 in., 3 in.

4. 7 ft

★ 5. 7 m, 7 m, 4 m, 5 m

Hint: Think of these shapes as more than 1 rectangle.

★ 6. 4 cm, 6 cm

PROBLEM SOLVING

7. Kevin's room measures 10 feet long and 12 feet wide. What is the area of his room?

8. A football field is 100 yards long and 50 yards wide. What is the area? What is the perimeter?

9. If the cover of a social studies book is 14 inches long and 7 inches wide, what is its area?

10. Henry's kitchen table is 6 feet long and 4 feet wide. What is the area of the table?

11. Kate lives 2 km from the store. She walked 900 m toward it. How many more meters must she walk to reach the store?

★ 12. Draw as many rectangles as you can that have an area of 20 square units. How many can you draw? (Use only whole numbers.)

MID-CHAPTER REVIEW

Identify the line segments as *parallel, intersecting,* or *perpendicular*.

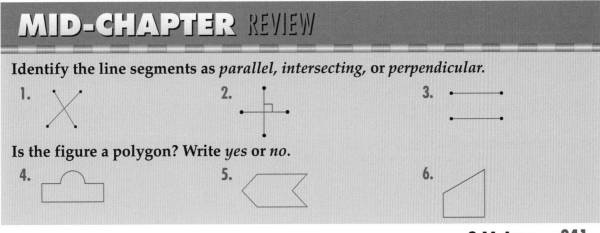

1.

2.

3.

Is the figure a polygon? Write *yes* or *no*.

4.

5.

6.

9.12 Solid Figures and Nets

Objective: to identify three-dimensional shapes; to identify a three-dimensional shape based on a two-dimensional representation

Thomas has a toy box. It has the shape of a rectangular prism. Each flat surface is called a **face**. Two faces meet at an **edge**. Three edges meet at a **vertex**.

Examples

Hold a model of a pyramid in your hand. Look at it from the top, sides, and bottom. Count the faces, edges, and vertices.

A. From the top, you can see the 8 edges and 5 vertices.

B. From the side you can see one of the triangular faces. This is a **square pyramid**.

C. From the bottom, you can see only the square face.

There are many three-dimensional figures.

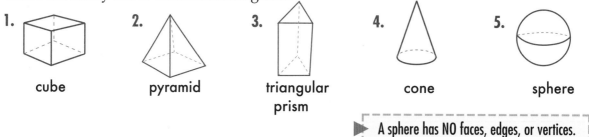

1. cube
2. pyramid
3. triangular prism
4. cone
5. sphere

▶ A sphere has NO faces, edges, or vertices.

If you were to unfold the first four three-dimensional shapes above, you would have a **net**, or a map of the shape. A **net** is a two-dimensional pattern of a three-dimensional solid.

This is the net of a cube.

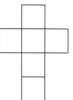

This is the net of a triangular pyramid.

TRY THESE

Choose the lettered figures that are parts of the solid figures shown.

1. cylinder

2. rectangular prism

3. pyramid

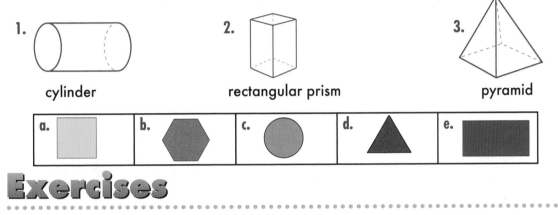

a.   b.   c.   d.   e.

Exercises

Complete the chart. Write the number of faces, edges, and vertices.

	Solid Figure	Faces	Edges	Vertices
1.	Cube			
2.	Rectangular prism			
3.	Pyramid			
4.	Triangular prism			

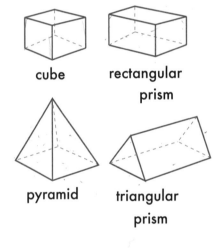

cube   rectangular prism

pyramid   triangular prism

PROBLEM SOLVING

Study the following figures and nets. Match each solid figure to its two-dimensional net.

5.   **6.**

7.   **8.**

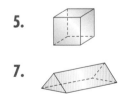

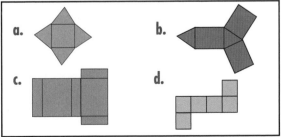

a.   b.   c.   d.

9.13 Volume

Objective: to find the volume of three-dimensional shapes

Volume is the total amount of space a figure takes up. The volume of a three-dimensional shape, like a cube or a rectangular prism, is determined by the length, width, and height. To find the volume, you multiply the three dimensions together.

Volume is measured in cubic units (units^3) because there are three dimensions. It is a measure of how many individual cubic units would fit inside the rectangular prism side-by-side until completely filled.

$$\text{length} \times \text{width} \times \text{height} = \text{Volume}$$
$$l \times w \times h = V$$

Example

Find the volume of the rectangular prism.

length = 7 inches
width = 6 inches
height = 5 inches

length × width × height = Volume
$$7 \quad \times \quad 6 \quad \times \quad 5 \quad = \quad 210$$
Volume = 210 cubic inches

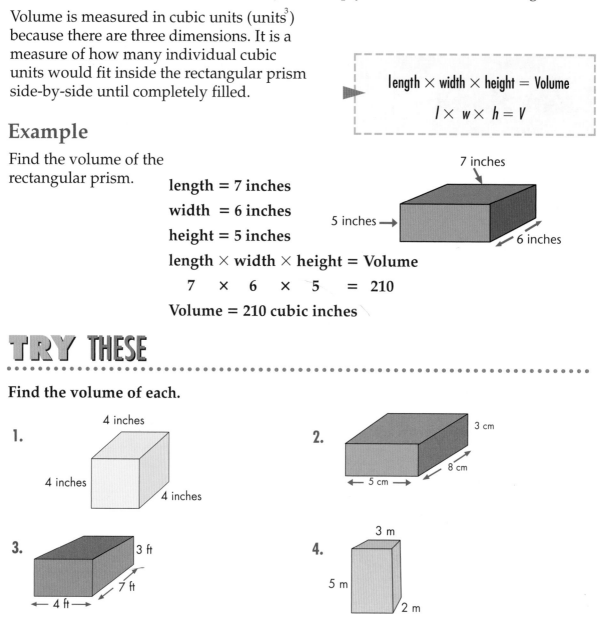

TRY THESE

Find the volume of each.

1.
4 inches
4 inches
4 inches

2.
3 cm
8 cm
5 cm

3.
3 ft
7 ft
4 ft

4.
3 m
5 m
2 m

Exercises

Find the volume of these three-dimensional shapes.

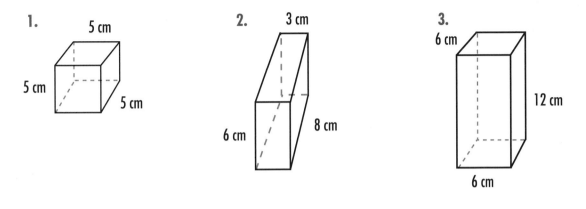

1.
5 cm
5 cm
5 cm

2.
3 cm
8 cm
6 cm

3.
6 cm
12 cm
6 cm

PROBLEM SOLVING

4. Inside the Great Pyramid in Egypt, the queen's chamber has a height of about 4 meters, a length of about 6 meters, and a width of about 5 meters. Find the approximate volume of the queen's chamber.

★**5.** The king's chamber in the Great Pyramid is much larger than the queen's chamber. Its volume is about 300 cubic meters. If the length of the chamber is 5 meters and the width is 10 meters, what is the height of the king's chamber?

★**6.** The Great Pyramid in Egypt has a length of 756 feet, a width of 756 feet, and a height of 481 feet. Find the volume of the pyramid by multiplying $l \times w \times h$, and then dividing by 3. (You need a different formula to find the volume of a pyramid than the one you use to find the volume of a rectangular prism or a cube because a pyramid takes up a different amount of space.)

9.14 Problem-Solving Strategy: Drawing a Diagram

Objective: to solve a problem by drawing a diagram or picture

Ella and Mia plan to make a large square table using 9 small tables pushed together in three rows of three tables. Each small table is 2 ft square. If they plan to tape a decorative border around the outside edge of the large table, how many feet will they need?

1. READ
Read the problem. You need to find out how much decorative border is needed to go around the outer edge of the large table. This requires knowing the perimeter of the large table.

2. PLAN
Draw a picture of the large table. Label the dimensions of the small table that make up the outer edges.

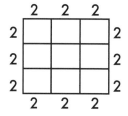

3. SOLVE
The perimeter is the distance around the outside of a polygon. In this case it will be 2 + 2 + 2 + 2 + 2 + 2 + 2 + 2 + 2 + 2 + 2 + 2 = 24 ft.

The girls will need 24 ft of border.

4. CHECK
You could use multiplication to check. There are twelve edges of the smaller tables that make the outer edge of the larger table.

$$12 \times 2 = 24 \text{ ft}$$

TRY THESE

Solve by drawing a picture.

1. Ms. Vlas has 144 square feet of carpet. Does she have enough carpet to cover a floor that measures 12 feet by 11 feet?

2. A meeting room has 4 rows of chairs. Each row has 14 chairs. How many chairs are in the room?

Solve

1. Mr. Black is planting a garden in 13-foot long rows. He places tomato plants 2 feet apart. How many plants will there be in two rows?

2. Mr. Black's garden is 14 feet long and 8 feet wide. He wants to put a fence around the garden. If he puts a stake every 2 feet, how many stakes will he need?

3. Mr. Leath is building a fence around his yard. The yard is a rectangle 30 yd by 50 yd. He has built 95 yd of fence. How much more fence must he build? Which measurement will help to solve—perimeter or area?

4. There are 25 students in a class. If there are the same number of students at each table and more than one table, how many tables are there?

5. Mrs. Connors wants to tile her kitchen floor. The floor is 12 feet long and 14 feet wide. Each tile measures 1 square foot. What is the area of her floor? How many tiles does she need?

6. Robert has a treehouse that measures 4 feet by 8 feet. What is the area of the treehouse?

CONSTRUCTED RESPONSE

7. You have three containers—Can A, Can B, and Can C. Can A holds 90 mL; Can B holds less than Can A; and Can C holds more than Can A. Which can holds the least amount? Which can holds the most amount? Explain your reasoning.

CHAPTER 9 REVIEW

LANGUAGE and CONCEPTS

Choose the correct word to complete each sentence.

1. Lines in the same plane that *never* cross are called (parallel, perpendicular) lines.

2. Lines in the same plane that cross are called (acute, intersecting) lines.

3. A closed plane figure with straight sides that do not cross is called a (polygon, pattern).

4. A four-sided polygon is called a (vertex, quadrilateral).

5. A five-sided polygon has five (angles, faces).

6. A figure that can be folded so that both sides match exactly has (line symmetry, right angles).

7. Two faces of a solid figure meet at an (obtuse, edge).

8. A solid figure that has six square faces is a (cube, cone).

SKILLS and PROBLEM SOLVING

Identify the line segments as *parallel, intersecting,* or *perpendicular.* (Section 9.2)

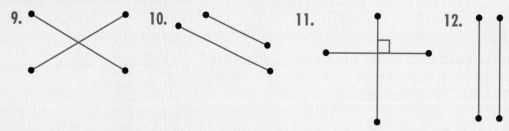

9.   **10.**   **11.**   **12.**

Compare each angle with a corner of your paper. Identify the angle as *right, acute,* or *obtuse.* (Section 9.3)

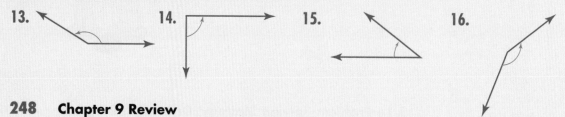

13.   **14.**   **15.**   **16.**

Name each polygon. How many lines of symmetry does each figure have? (Sections 9.4–9.5 and 9.8)

17.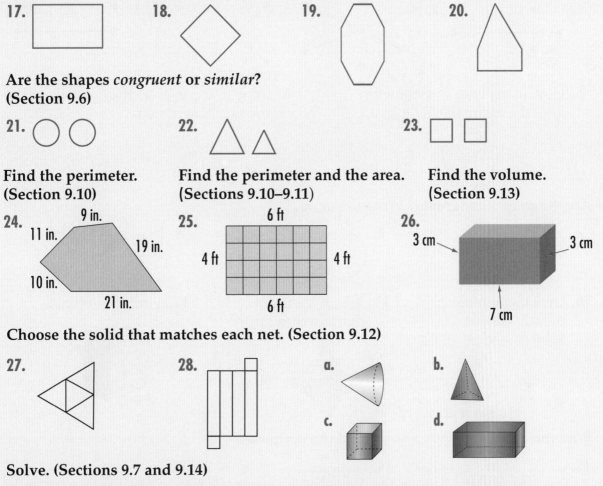

18.

19.

20.

Are the shapes *congruent* or *similar*? (Section 9.6)

21.

22.

23.

Find the perimeter. (Section 9.10)

Find the perimeter and the area. (Sections 9.10–9.11)

Find the volume. (Section 9.13)

24. 9 in. 11 in. 19 in. 10 in. 21 in.

25. 6 ft 4 ft 4 ft 6 ft

26. 3 cm 3 cm 7 cm

Choose the solid that matches each net. (Section 9.12)

27.

28.

a.

b.

c.

d.

Solve. (Sections 9.7 and 9.14)

29. Mrs. Kalb wants to plant a hedge along a 16-foot driveway. The plants need to be 2 feet apart with 1 foot of space at either end. How many plants does she need to buy? Draw a picture to help you solve.

30. Which one does not belong? Give a reason.

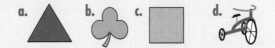

a. b. c. d.

31. Kate's living room is 30 feet wide, 45 feet long, and 10 feet tall. What is the volume of Kate's living room?

32. Amy wants to plant flowers around her house. The length of the house is 75 feet, and the width is 89 feet. What is the perimeter of Amy's house?

33. Justin has a large box of books. Draw what the net would look like for the box.

Identify the line segments as *parallel, intersecting,* or *perpendicular.*

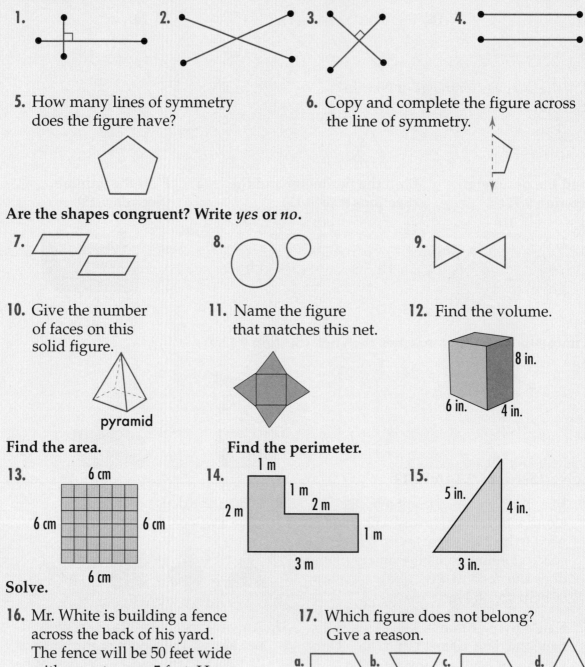

1.

2.

3.

4.

5. How many lines of symmetry does the figure have?

6. Copy and complete the figure across the line of symmetry.

Are the shapes congruent? Write *yes* or *no*.

7.

8.

9.

10. Give the number of faces on this solid figure.

pyramid

11. Name the figure that matches this net.

12. Find the volume.

8 in.

6 in.   4 in.

Find the area.

13.   6 cm

6 cm   6 cm

6 cm

Find the perimeter.

14.   1 m

1 m

2 m   2 m

1 m

3 m

15.

5 in.   4 in.

3 in.

Solve.

16. Mr. White is building a fence across the back of his yard. The fence will be 50 feet wide with a post every 5 feet. How many posts will he need?

17. Which figure does not belong? Give a reason.

a.   b.   c.   d.

CHANGE OF PACE

Nine Men's Morris

Possible Game Pieces
pennies
buttons
counters
beads
small tiles

"Nine Men's Morris" is a game of strategy, similar to tic-tac-toe but with twists. Where it began, no one really knows, but it is believed that the ancient Egyptians played this game. It is a game for two players. Each player needs nine game pieces, or "men," that will fit on the following game board.

The goal of the game is to capture enough of your opponent's men so that he is unable to get three in a row. Another way to win is to trap your opponent, leaving him with no available moves.

There are two phases for the game:

Phase 1: Laying Pieces on the Board

Play begins with one empty game board. Each player takes a turn placing one game piece on any of the game board stars. When a player has three pieces in a row, he has a "mill." The mill must be straight; it cannot turn a corner; and it must be along a drawn line. Once a player has a mill, he may remove any one of his opponent's pieces from the board. A piece may be part of more than one mill at a time.

Phase 2: Continuing Forth

After all eighteen pieces have been played, the game continues by having each player take turns sliding one man along any drawn line to the next open space. Players keep looking for opportunities to make mills to capture another of their opponent's men. There are no jumps in this game, so if another man is in the way, another move must be found. Play continues until one player has only two pieces or no open moves.

Now it is your turn to try your hand at "Nine Men's Morris."

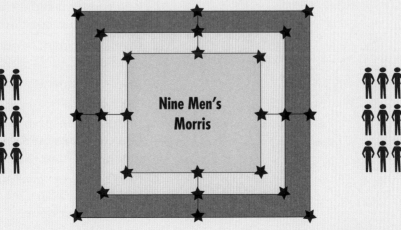

CUMULATIVE TEST

1. 3)533
   a. 111 R2
   b. 177 R2
   c. 166 R2
   d. none of the above

2. What is another way to give the time 6:45?
   a. a quarter after 7
   b. 15 minutes after 6
   c. 15 minutes before 7
   d. none of the above

3. How many faces does a cube have?
   a. 6 faces
   b. 8 faces
   c. 12 faces
   d. none of the above

4. What word describes the relationship of these line segments?
   a. intersecting
   b. parallel
   c. perpendicular
   d. none of the above

5. $27.32
   − 19.58
   a. $7.74
   b. $8.74
   c. $12.26
   d. none of the above

6. 285
   × 60
   a. 1,710
   b. 17,100
   c. 20,430
   d. none of the above

7. Mrs. Chu Ling wants to buy 100 pepper plants. The pepper plants are sold in trays of 6 plants. How many trays does she need to buy?
   a. 16 trays
   b. 17 trays
   c. 600 trays
   d. none of the above

8. Which figure is a polygon?
   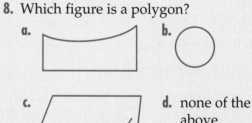
   d. none of the above

9. Which figure does not belong?
   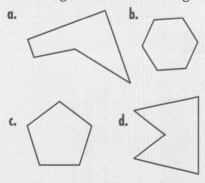

10. Jack babysat for 7 hours. Lita worked at the record shop for 5 hours. Juan said that Jack earned more money than Lita. Which of these is correct about Juan's statement?
    a. Juan's statement is true under all conditions.
    b. Juan's statement cannot be true under any conditions.
    c. Juan's statement is true if Jack and Lita each earn the same amount of money per hour.
    d. Juan's statement cannot be true if Jack and Lita each earn the same amount of money per hour.

CHAPTER 10

Fractions and Probability

Forrest Northey
Florida

10.1 Representing Fractions

Objective: to represent fractions as part of a whole or part of a set; to write fractions in correct form and in words

Each summer, the Smitson children make craft projects for the county fair. Lucy is making a design with felt. Each triangle of felt is one-fourth of a square piece. You can write one-fourth as the **fraction** $\frac{1}{4}$.

1 out of 4
one-fourth

$\frac{1}{4}$ → **numerator**: the number of parts you are talking about
 → **denominator**: the number of equal-sized parts in the whole

Next, Lucy took one of the four bottles of glue to glue down her felt triangle. She is using $\frac{1}{4}$ of the glue she bought.

With the felt, $\frac{1}{4}$ represents 1 out of 4 parts of a whole. With the glue, $\frac{1}{4}$ represents 1 out of a set of 4 objects.

More Examples

A. Two-thirds of the flag is green.

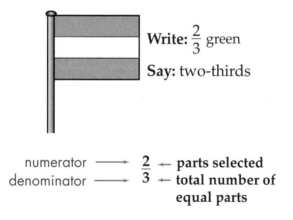

Write: $\frac{2}{3}$ green

Say: two-thirds

numerator ⟶ $\frac{2}{3}$ ← **parts selected**
denominator ⟶ ← **total number of equal parts**

B. Three-fifths of the balls are blue

numerator ⟶ $\frac{3}{5}$ ← **parts selected**
denominator ⟶ ← **total number of objects in the set**

TRY THESE

Is the figure divided into equal parts? Write *yes* or *no*. Write a fraction for the colored part of each figure that has been divided into equal parts.

1. 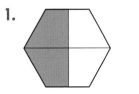　**2.** 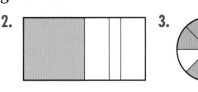　**3.** 　**4.**

Solve.

5. What fraction of the flowers are white?

6. What fraction of the tools are hammers?

7. What fraction of the balls of yarn are green?

Exercises

Write a fraction for the colored part of each figure or set. Then write the fraction in words.

1.　**2.**　**3.**　**4.**

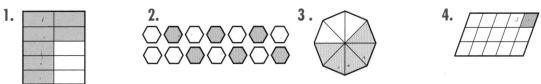

Draw two pictures for each fraction: as a part of a whole and as a part of a set.

5. $\dfrac{3}{10}$　　**6.** $\dfrac{2}{5}$　　**7.** $\dfrac{7}{8}$

PROBLEM SOLVING

8. Rosa bought a set of eight markers. She lost three markers. What fraction of the set is lost?

10. How many tomato plants are there in 8 trays like this one?

9. Jennifer brought home ten books from the library. She reads three of them. What fraction of the books does she still have to read?

11. What fraction of the pizza has been eaten?

10.2 Fraction of a Number

Objective: to find the fractional part of a number using visual models

Toby and Jane picked apples. Toby gave $\frac{1}{3}$ of his 6 apples to Jane.
How many apples did he give her?

The denominator, 3, tells you to separate the apples into three groups of the same number.

$$\frac{1}{3} \text{ of 6 apples} = 2 \text{ apples}$$

$$\frac{1}{3} \text{ of } 6 = 2$$

Example

John Poling will use $\frac{2}{3}$ of a dozen eggs to make breakfast
for his family. How many eggs will he use? Find $\frac{2}{3}$ of 12.

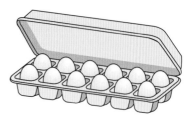

THINK   I know how to find one-third of 12.
Two-thirds would be twice as much.

Step 1	Step 2
Find one-third of 12. 1 of 3 equal parts	Find two-thirds of 12. 2 of 3 equal parts
One-third of 12 eggs is 4 eggs.	Two-thirds of 12 eggs is 8 eggs.
$\frac{1}{3}$ of 12 is 4.	$\frac{2}{3}$ of 12 is 8.

John will use 8 eggs.

More Examples

A. Find $\frac{1}{4}$ of 24.

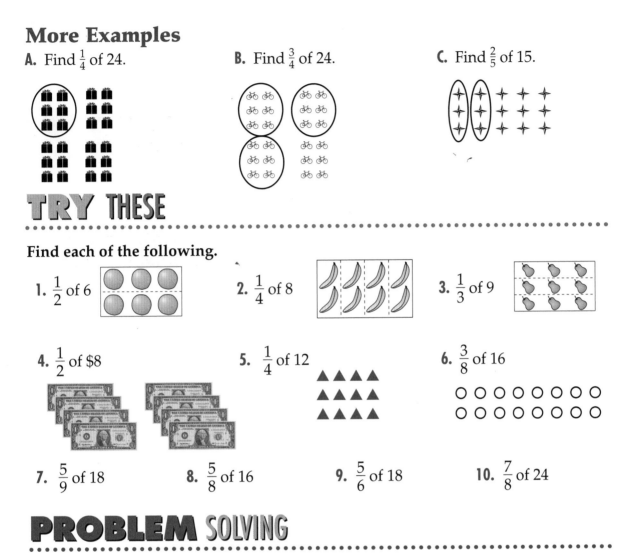

B. Find $\frac{3}{4}$ of 24.

C. Find $\frac{2}{5}$ of 15.

TRY THESE

Find each of the following.

1. $\frac{1}{2}$ of 6

2. $\frac{1}{4}$ of 8

3. $\frac{1}{3}$ of 9

4. $\frac{1}{2}$ of $8

5. $\frac{1}{4}$ of 12

6. $\frac{3}{8}$ of 16

7. $\frac{5}{9}$ of 18

8. $\frac{5}{8}$ of 16

9. $\frac{5}{6}$ of 18

10. $\frac{7}{8}$ of 24

PROBLEM SOLVING

Use the chart to solve.

11. Four-fifths of the lifesaving class are adults. How many adults are taking lifesaving?

12. Five-ninths of the swimming students are under 12 years old. How many swimming students are under 12 years old?

13. One-third of the diving students are in an evening class. How many diving students are in an evening class?

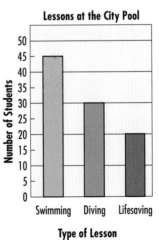

Lessons at the City Pool

10.3 Equivalent Fractions

Objective: to find equivalent fractions using models

Robin marked a board into sixths. She painted two sections or $\frac{2}{6}$ of the board.

Arnold marked a board of the same length into thirds. Then he painted $\frac{1}{3}$ of the board.

Robin and Arnold painted the same amount.

Fractions like $\frac{2}{6}$ and $\frac{1}{3}$ that name the same amount are **equivalent fractions**.

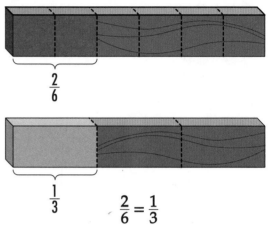

$$\frac{2}{6} = \frac{1}{3}$$

More Examples

To find equivalent fractions, multiply or divide both the numerator and denominator by the same number.

A. $\frac{2}{3} = \frac{\blacksquare}{6}$

$\frac{2 \times 2}{3 \times 2} = \frac{\blacksquare}{6} = \frac{4}{6}$

so $\frac{2}{3} = \frac{4}{6}$

B. $\frac{4}{6} = \frac{\blacksquare}{3}$

$\frac{4 \div 2}{6 \div 2} = \frac{\blacksquare}{3} = \frac{2}{3}$

so $\frac{4}{6} = \frac{2}{3}$

TRY THESE

Find equivalent fractions.

1. $\frac{1}{2} \times \frac{2}{2} = \frac{\blacksquare}{4}$

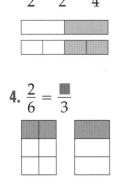

2. $\frac{1}{4} = \frac{2}{\blacksquare}$

3. $\frac{10}{16} = \frac{\blacksquare}{8}$

4. $\frac{2}{6} = \frac{\blacksquare}{3}$

5. $\frac{4}{5} = \frac{\blacksquare}{10}$

6. $\frac{3}{4} = \frac{6}{\blacksquare}$

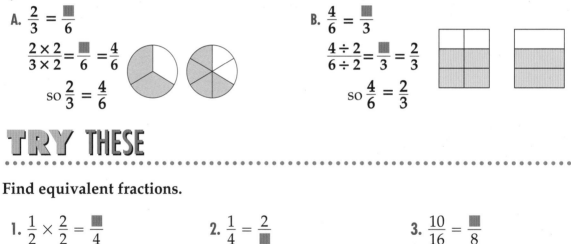

Exercises

Choose two equivalent fractions for the colored part.

1.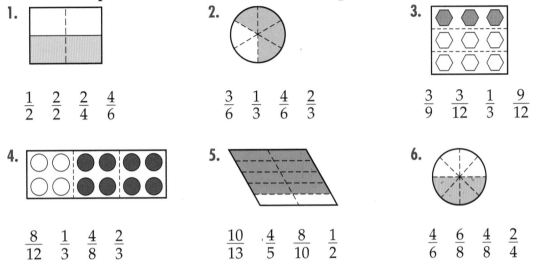

$\frac{1}{2}$ $\frac{2}{2}$ $\frac{2}{4}$ $\frac{4}{6}$

2.

$\frac{3}{6}$ $\frac{1}{3}$ $\frac{4}{6}$ $\frac{2}{3}$

3.

$\frac{3}{9}$ $\frac{3}{12}$ $\frac{1}{3}$ $\frac{9}{12}$

4.

$\frac{8}{12}$ $\frac{1}{3}$ $\frac{4}{8}$ $\frac{2}{3}$

5.

$\frac{10}{13}$ $\frac{4}{5}$ $\frac{8}{10}$ $\frac{1}{2}$

6.

$\frac{4}{6}$ $\frac{6}{8}$ $\frac{4}{8}$ $\frac{2}{4}$

Replace each ▊ with a number so the fractions are equivalent. Use fraction models.

7. $\frac{1 \times 2}{4 \times 2} = \frac{▊}{8}$

8. $\frac{1 \times 3}{2 \times 3} = \frac{3}{▊}$

9. $\frac{1 \times 2}{5 \times 2} = \frac{2}{▊}$

10. $\frac{1 \times 3}{3 \times 3} = \frac{▊}{9}$

PROBLEM SOLVING

11. Name two equivalent fractions to show how much of this carton is filled with eggs.

12. Linda raked $\frac{1}{2}$ of the yard. How many fourths did she rake?

13. Ned ate $\frac{3}{4}$ of the pack of cookies. How many sixteenths did he eat?

TEST PREP

14. Five-eighths of the fish that David caught were bluefish. How many twenty-fourths of his catch were bluefish?

a. $\frac{10}{24}$ b. $\frac{15}{24}$ c. $\frac{18}{24}$ d. $\frac{20}{24}$

10.4 Fractions in Simplest Form

Objective: to reduce and write fractions in simplest form

Twelve of eighteen walnuts were eaten by raccoons.

Write the simplest form of $\frac{12}{18}$.

First, you need to find the **greatest common factor** of 12 and 18.

Factors of 12: 1, 2, 3, 4, 6, 12

Factors of 18: 1, 2, 3, 6, 9, 18

The greatest factor they share is 6. Divide the numerator and denominator by 6.

 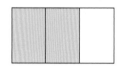

$$\frac{12}{18} = \frac{12 \div 6}{18 \div 6} = \frac{2}{3} \qquad \frac{12}{18} \text{ in simplest form is } \frac{2}{3}.$$

> A fraction is in **simplest form** when the greatest common factor that the numerator and denominator share is 1.

More Examples

A. $\frac{8}{10} = \frac{\blacksquare}{5} \cdots \cdots \frac{8}{10} = \frac{4}{5}$

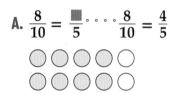

B. $\frac{12}{16} = \frac{\blacksquare}{\blacksquare} \cdots \cdots \frac{12}{16} = \frac{3}{4}$

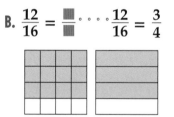

TRY THESE

Copy and complete.

1. $\frac{2}{4} = \frac{\blacksquare}{2}$

2. $\frac{6}{9} = \frac{\blacksquare}{\blacksquare}$

3. $\frac{4}{12} = \frac{\blacksquare}{\blacksquare}$

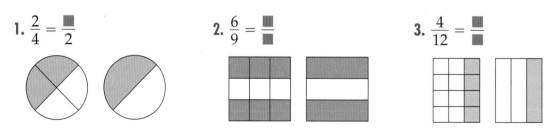

Exercises

Write an equivalent fraction in simplest form.

1. $\dfrac{3 \div 3}{9 \div 3} = \blacksquare$

2. $\dfrac{2 \div 2}{4 \div 2} = \blacksquare$

3. $\dfrac{6 \div 2}{14 \div 2} = \blacksquare$

4. $\dfrac{10 \div 5}{15 \div 5} = \blacksquare$

5. $\dfrac{6}{8}$

6. $\dfrac{8}{10}$

7. $\dfrac{6}{15}$

8. $\dfrac{4}{8}$

9. $\dfrac{4}{12}$

10. $\dfrac{15}{20}$

11. $\dfrac{4}{6}$

12. $\dfrac{3}{21}$

13. $\dfrac{20}{30}$

14. $\dfrac{12}{16}$

15. $\dfrac{7}{28}$

16. $\dfrac{10}{12}$

PROBLEM SOLVING

17. Tai has 237 cucumbers to put in jars for pickles. He can put an average of 7 cucumbers in a jar. How many jars does Tai need?

18. What fraction of the marbles are *not* blue?

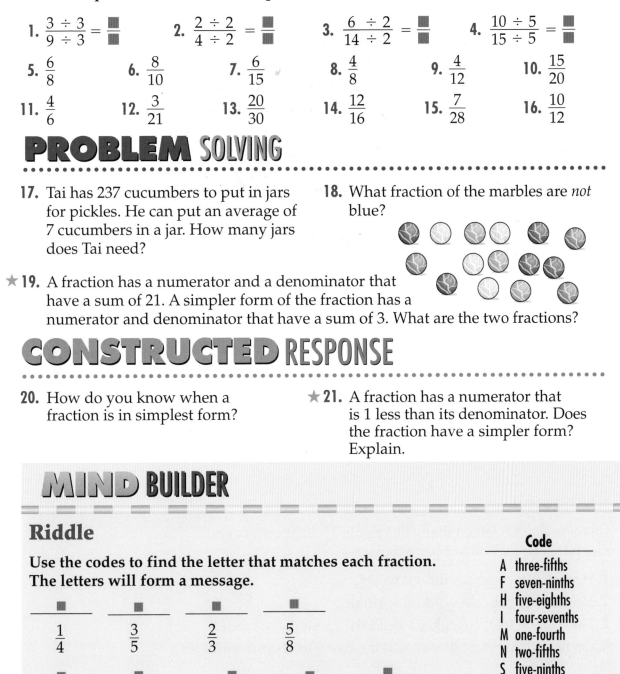

★ 19. A fraction has a numerator and a denominator that have a sum of 21. A simpler form of the fraction has a numerator and denominator that have a sum of 3. What are the two fractions?

CONSTRUCTED RESPONSE

20. How do you know when a fraction is in simplest form?

★ 21. A fraction has a numerator that is 1 less than its denominator. Does the fraction have a simpler form? Explain.

MIND BUILDER

Riddle

Use the codes to find the letter that matches each fraction. The letters will form a message.

Code	
A	three-fifths
F	seven-ninths
H	five-eighths
I	four-sevenths
M	one-fourth
N	two-fifths
S	five-ninths
T	two-thirds
U	one-half

$\dfrac{1}{4}$ $\dfrac{3}{5}$ $\dfrac{2}{3}$ $\dfrac{5}{8}$

$\dfrac{4}{7}$ $\dfrac{5}{9}$ $\dfrac{7}{9}$ $\dfrac{1}{2}$ $\dfrac{2}{5}$.

Problem Solving

Counting Cupcakes

Matt, Katt, Bill, and Jill were having a baking bonanza one Saturday. They baked 90 cupcakes among them. Matt made 2 dozen cupcakes. He put sprinkles on top of $\frac{1}{4}$ of his cupcakes. Bill made 2 less cupcakes than Matt. He left $\frac{1}{2}$ of his without sprinkles. Jill made $1\frac{2}{3}$ dozen cupcakes. She sprinkled $2\frac{1}{5}$ of her cupcakes. Katt made 4 more cupcakes than Jill. She put sprinkles on only $\frac{1}{6}$ of hers. At the end of the baking bonanza, on how many cupcakes total did the four friends put sprinkles?

Extension

Imagine that Lil joined them. She made $1\frac{3}{4}$ dozen cupcakes. She put sprinkles on $\frac{1}{3}$ of her cupcakes.

1. How many cupcakes did Lil bake?

2. How many cupcakes did Lil sprinkle?

3. How many total cupcakes would the group have made?

4. On how many cupcakes would the five friends put sprinkles?

CUMULATIVE REVIEW

Multiply or divide.

1. 357×6
    2. 24×58
    3. $15 \times \$8.21$

4. $7\overline{)68}$
    5. $4\overline{)72}$
    6. $3\overline{)463}$

7. $2,000 \times 7$
    8. $5\overline{)89}$
    9. $7 \times 6 \times 2$

Estimate.

10. $\begin{array}{r} 864 \\ \times\ 8 \\ \hline \end{array}$
  11. $\begin{array}{r} 156 \\ +\ 728 \\ \hline \end{array}$
  12. $\begin{array}{r} 52 \\ \times\ 14 \\ \hline \end{array}$
  13. $\begin{array}{r} 475 \\ \times\ 42 \\ \hline \end{array}$
  14. $\begin{array}{r} 3,674 \\ -\ 2,704 \\ \hline \end{array}$

Name the unit that is best to measure each of the following.
Write _meter, centimeter, gram, kilogram, liter,_ or _milliliter._

15. the length of a nail
    16. water in a pitcher

17. a bag of flour
    18. the width of a closet

19. soup in a spoon
    20. the mass of a car

Give the number of faces on each solid figure.

21.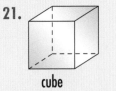
cube

22.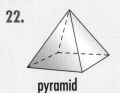
pyramid

23.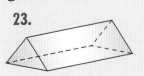
triangular prism

★**24.**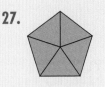
pentagonal prism

Write a fraction for each colored part.

25.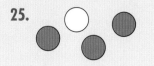
    26.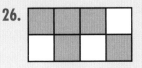
    27.

Solve.

28. Juan has 57 baseball cards. Sue has 3 times as many cards as Juan. How many baseball cards does Sue have?

29. Sam has 2 quarters, 7 dimes, and 8 pennies. How many 7¢ pencils can he buy? How much money will he have left?

10.5 Comparing and Ordering Fractions

Objective: to compare and order fractions using models

Tonya wants to paint the doghouse. She has $\frac{3}{8}$ of a jar of red paint and $\frac{5}{8}$ of a jar of blue paint. Does she have more red or blue paint?

To compare fractions with the same denominator, compare the numerators.

Since $5 > 3$, then $\frac{5}{8} > \frac{3}{8}$.

So, Tonya has more blue paint than red paint.

THINK

On the number line, $\frac{5}{8}$ is to the right of $\frac{3}{8}$. So, $\frac{5}{8} > \frac{3}{8}$.

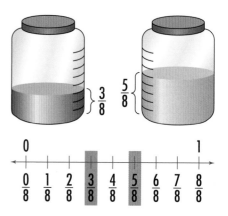

More Examples

Use pictures of equivalent fractions to help compare fractions with different denominators.

A. $\frac{1}{4} \bullet \frac{1}{8}$

B. $\frac{2}{3} \bullet \frac{5}{6}$

C. $\frac{1}{2} \bullet \frac{3}{8}$

$\frac{1}{4} > \frac{1}{8}$

$\frac{2}{3} < \frac{5}{6}$

$\frac{1}{2} > \frac{3}{8}$

TRY THESE

Compare using <, >, or =.

1. $\frac{4}{6} \bullet \frac{6}{6}$

2. $\frac{2}{4} \bullet \frac{3}{4}$

3. $\frac{7}{9} \bullet \frac{4}{9}$

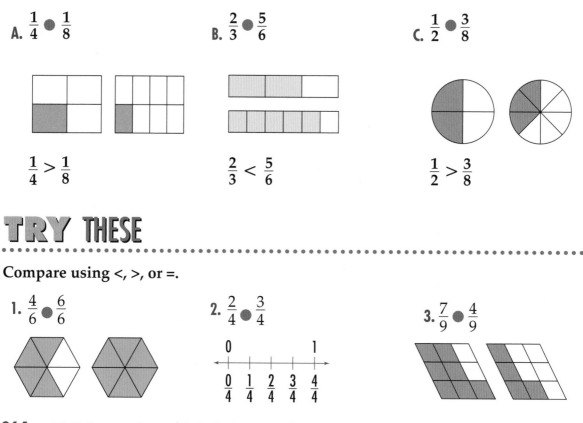

Exercises

Compare using <, >, or =.

1. $\dfrac{5}{8}$ ● $\dfrac{1}{2}$

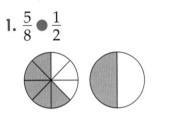

2. $\dfrac{1}{3}$ ● $\dfrac{5}{12}$

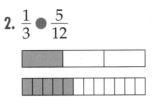

3. $\dfrac{2}{6}$ ● $\dfrac{1}{3}$

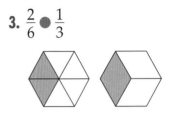

4. $\dfrac{7}{12}$ ● $\dfrac{2}{3}$

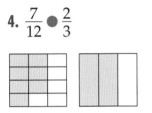

5. $\dfrac{7}{8}$ ● $\dfrac{5}{8}$

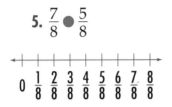

6. $\dfrac{3}{4}$ ● $\dfrac{3}{12}$

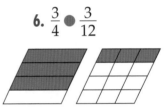

Use the fraction strips on the right to compare each pair of fractions. Use <, >, or =.

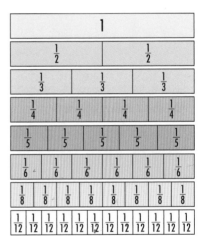

7. $\dfrac{2}{3}$ ● $\dfrac{3}{4}$

8. $\dfrac{1}{6}$ ● $\dfrac{1}{4}$

9. $\dfrac{1}{2}$ ● $\dfrac{3}{8}$

10. $\dfrac{1}{6}$ ● $\dfrac{1}{8}$

11. $\dfrac{2}{8}$ ● $\dfrac{3}{12}$

12. $\dfrac{3}{6}$ ● $\dfrac{3}{4}$

13. $\dfrac{2}{4}$ ● $\dfrac{3}{6}$

14. $\dfrac{2}{8}$ ● $\dfrac{2}{6}$

15. $\dfrac{7}{8}$ ● $\dfrac{6}{12}$

PROBLEM SOLVING

16. One nail is $\dfrac{9}{16}$ of an inch long. Another nail is $\dfrac{5}{8}$ of an inch long. Which nail is longer?

17. Mrs. Heath had $36 in her wallet. She spent $\dfrac{1}{2}$ of her money on a toaster. How much did the toaster cost?

★ 18. Circles cost 10¢; squares cost 5¢; and triangles cost 2¢. What is the cost of the picture on the right? Make a picture that costs 31¢.

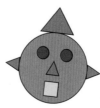

10.6 Mixed Numbers and Improper Fractions

Objective: to represent fractions greater than one whole as mixed numbers and as improper fractions

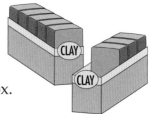

When Jessica finishes her chores, she enjoys making things with clay. How many boxes of clay does Jessica have?

Jessica counts 8 blocks of clay. There are 5 blocks of clay in a full box.

$\frac{8}{5}$ $8 > 5$ The numerator is greater than the denominator.

$\frac{8}{5}$ is an **improper fraction**. An improper fraction names a number greater than or equal to 1.

An improper fraction also can be written as a mixed number.
A **mixed number**, such as $1\frac{3}{5}$, has a whole number part and a fraction part.

$1\frac{3}{5} = 1 + \frac{3}{5}$ one and three-fifths

Jessica has $1\frac{3}{5}$ boxes of clay.

More Examples

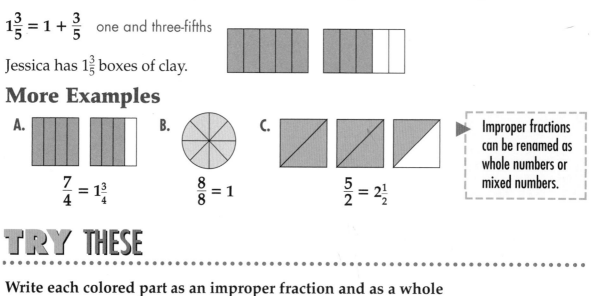

A. $\frac{7}{4} = 1\frac{3}{4}$

B. $\frac{8}{8} = 1$

C. $\frac{5}{2} = 2\frac{1}{2}$

▶ Improper fractions can be renamed as whole numbers or mixed numbers.

TRY THESE

Write each colored part as an improper fraction and as a whole number or mixed number.

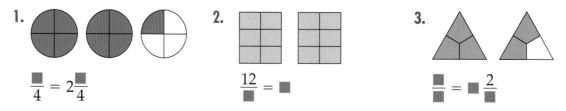

1. $\frac{\blacksquare}{4} = 2\frac{\blacksquare}{4}$

2. $\frac{12}{\blacksquare} = \blacksquare$

3. $\frac{\blacksquare}{\blacksquare} = \blacksquare\frac{2}{\blacksquare}$

Exercises

Write each colored part as an improper fraction and as a whole number or mixed number.

1.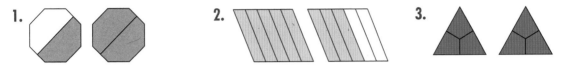

2.

3.

Rename each improper fraction as a whole number or a mixed number. Make or draw a model to show your answer.

4. $\frac{12}{10}$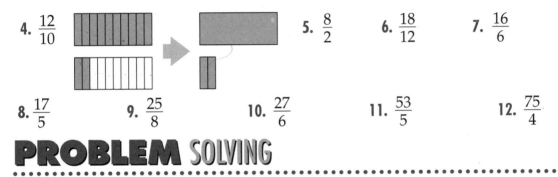

5. $\frac{8}{2}$

6. $\frac{18}{12}$

7. $\frac{16}{6}$

8. $\frac{17}{5}$

9. $\frac{25}{8}$

10. $\frac{27}{6}$

11. $\frac{53}{5}$

12. $\frac{75}{4}$

PROBLEM SOLVING

★ **13.** Nora has 3 pizzas. She slices each pizza into thirds. After the meal, she has 7 pieces left over. What is this as a mixed number?

14. Hank cut some boards into halves. He used ten halves. How many whole boards did he use?

MIND BUILDER

Equivalent Fractions

You can find equivalent fractions on a number line. For example, $\frac{1}{4}$ is equivalent to $\frac{2}{8}$.

$$\frac{1}{4} = \frac{2}{8}$$

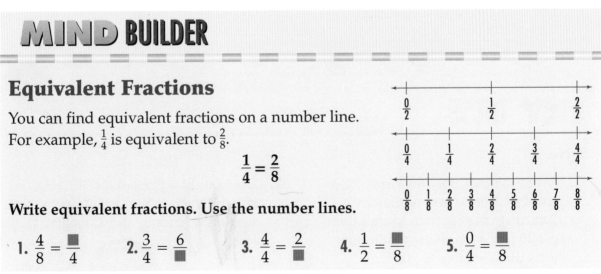

Write equivalent fractions. Use the number lines.

1. $\frac{4}{8} = \frac{\blacksquare}{4}$

2. $\frac{3}{4} = \frac{6}{\blacksquare}$

3. $\frac{4}{4} = \frac{2}{\blacksquare}$

4. $\frac{1}{2} = \frac{\blacksquare}{8}$

5. $\frac{0}{4} = \frac{\blacksquare}{8}$

10.7 Problem-Solving Strategy: Working Backwards

Objective: to solve problems by working backwards

The pies for a picnic were cut into sixths. There were 18 pieces of pie. How many whole pies were cut into pieces?

This problem tells the result and asks for something that happened earlier. The plan of working backwards can be used to solve this problem.

| **1. READ** | How many whole pies were cut into pieces? Each pie was cut into 6 pieces, and there were 18 pieces. |

| **2. PLAN** | Work backwards. If there were 18 pieces when each pie was cut into sixths, how many 6s are in 18? Divide 18 by 6 to find the number of whole pies. |

| **3. SOLVE** | $\begin{array}{r} 3 \\ 6\overline{)18} \\ -18 \\ \hline 0 \end{array}$ There were 3 whole pies cut into pieces. |

| **4. CHECK** | Three pies cut into 6 pieces each would give a total of 18 pieces. The answer checks. |

$6 \times 3 = 18$

TRY THESE

Solve.

1. Margo made some note cards. She gave 24 cards to her mother and 12 cards to her teacher. Margo then had 29 cards left. How many note cards did she have to start?

2. Luis bought a pretzel for 35¢. This is how much change the clerk gave him. How much money did Luis give the clerk?

Solve

Use any strategy to solve.

1. Harvey had a long piece of wire. He cut off two 8-inch pieces. Then he had 25 inches of wire left. How long was the wire before Harvey cut it?

2. Jackie had some stamps. She gave half of the stamps to Jan. She gave the remaining 6 stamps to Brad. How many stamps did Jackie give away?

3. Sue has 7 red balloons and 2 yellow balloons. What fraction of the balloons are yellow?

4. Eight children need to share four bananas equally. How much should each child receive?

5. Roberta bought 4 new books. Now she has 17 books altogether. How many books did she have to start?

6. Fred has 3 times as many quarters as dimes. He has 12 quarters. How many dimes does Fred have?

7. Esteban bought 6 pens for $0.25 each. He also bought a notebook for $1.75. How much money did Esteban start with if he had no money left?

8. The seal at Oceanland Park weighs 372 pounds. How much less does the seal weigh than a 744-pound, prize-winning bull at the county fair?

CONSTRUCTED RESPONSE

9. Dean wanted to bring enough popsicles to school for each person in his entire grade to have 1. The popsicles come 12 to a box. There are 71 students in his grade. How many boxes of popsicles will he need to buy? Explain your reasoning.

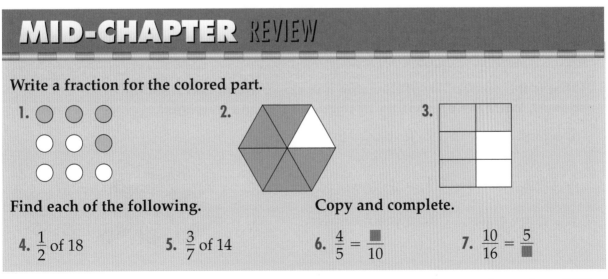

MID-CHAPTER REVIEW

Write a fraction for the colored part.

1.
2.
3.

Find each of the following.

4. $\frac{1}{2}$ of 18

5. $\frac{3}{7}$ of 14

Copy and complete.

6. $\frac{4}{5} = \frac{\blacksquare}{10}$

7. $\frac{10}{16} = \frac{5}{\blacksquare}$

10.8 Finding Probability

Objective: to find the probability of an event's outcome

Maria's father bought a variety pack of 6 fruit juices. Maria chooses one juice at random. To make the choice random, she must *not* look at the juices. **Random** means that each choice is equally likely.

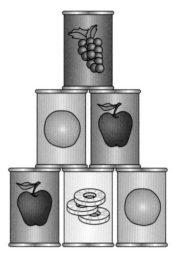

There are 6 possible outcomes. The chance, or **probability**, of choosing the grape juice is 1 out of 6, or $\frac{1}{6}$.

Another Example

A basket contains 3 white and 4 yellow tennis balls. What is the probability that a yellow tennis ball is chosen?

There are 7 possible outcomes. Four of the outcomes are yellow tennis balls. So, the probability is 4 out of 7, or $\frac{4}{7}$, that a yellow tennis ball is chosen.

TRY THESE

What is the probability of choosing one eraser at random from those shown? Write the probability in fraction form and in words.

1. a pink eraser
2. an eraser that is not pink
3. a blue eraser
4. any eraser
5. a green eraser
6. a yellow eraser

Exercises

Write the probability of choosing each of the following.

1. a $1.00 bill

2. a blue bill

3. a $100 bill

4. a yellow bill

5. a white or pink bill

Answer the following questions.

6. Is it possible to have a probability of 1? If so, what does it mean?

7. Is it possible to have a probability of 0? If so, what does it mean?

PROBLEM SOLVING

8. A number cube is rolled. How many possible outcomes are there? What is the probability of rolling an even number?

9. A reading group has 4 girls and 4 boys. One student is chosen at random. How many possible outcomes are there? What is the probability that a girl is chosen?

10. Dora spent $252 on some clay pots. Each cost $6. How many clay pots did she buy?

11. Jamie ate $\frac{1}{2}$ of a pizza. His sister ate $\frac{3}{8}$ of the pizza. Who ate more?

CONSTRUCTED RESPONSE

12. Explain how you would find $\frac{5}{7}$ of 35 without using a drawing or counters.

MIXED REVIEW

Find the average of each set of numbers.

13. 12, 15, 15, 18, 20

14. 26, 21, 45, 32

15. 64, 83, 51

10.9 Making Predictions

Objective: to make predictions of probability based on previous results

Katia has a spinner. The spinner is divided into six sections. Three sections are pink; one section is green; one section is blue; and one section is yellow. After spinning the spinner 12 times, it landed on the pink section 6 times. The spinner landed on pink 6 out of 12 times, or $\frac{6}{12}$. Katia knew that meant it landed on pink $\frac{1}{2}$ of the times. Katia started to make predictions (a reasonable guess) based on her experience. She wondered how many times the spinner would land on pink if she spun it 100 times. Based on what had already happened, she predicted it would land on pink 50 times because 50 is half of 100 times. The only way she would know for sure would be to test it out. But, based on her experience, her prediction was reasonable.

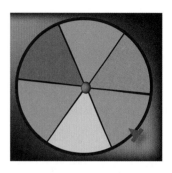

Another Example

Katia spun the spinner 20 times and twice it landed on blue. If she were to spin it 10 more times, how many more times will it land on blue?

> **THINK** It already landed on blue $\frac{2}{20}$ times. Since 10 is half of 20, a good prediction would have the spinner land on blue half the amount of times it already did. One-half of 2 is 1.

A prediction of $\frac{1}{10}$ more times on blue would be reasonable.

TRY THESE

Use the following information to answer each question.

John has a bag of 8 gumballs. In the bag, 4 gumballs are red, 2 are green, 1 is blue, and 1 is white.

1. What is the probability that John will pick a red gumball?

2. What is the probability of John picking a white gumball?

3. **Make a prediction**: If John were to try this experiment 40 times, about how many times would he pick out a green gumball?

Exercises

Imagine you are rolling a number cube 60 times. What is a prediction for each of the following? Write your answer in fraction form.

1. roll a 3

2. roll an even number

3. roll a 5 or a 6

PROBLEM SOLVING

4. A relay team has 3 girls and 5 boys as members. The coach is going to choose a captain by drawing one of the runners' names out of a hat. How many outcomes are there? **Make a prediction**: What is the probability that a boy is chosen?

5. There are 2 red and 4 green marbles in a bag. You chose 1 marble. How many outcomes are there? What is the probability of choosing a green marble?

CONSTRUCTED RESPONSE

6. Will your predicted probability for **Exercises 1–3** be identical to your actual probability after 60 rolls? Why or why not? Explain.

MIXED REVIEW

Divide.

7. $2\overline{)18}$

8. $3\overline{)27}$

9. $6\overline{)51}$

10. $5\overline{)57}$

11. $4\overline{)64}$

12. $7\overline{)203}$

13. $6\overline{)435}$

14. $4\overline{)812}$

15. $3\overline{)4,162}$

16. $8\overline{)3,679}$

17. $743 \div 6$

18. $505 \div 3$

19. $8,719 \div 4$

20. $6,347 \div 5$

10.10 Organizing Possible Outcomes

Objective: to use tree diagrams to organize outcomes

Kim eats a balanced lunch of milk, a sandwich, and a fruit. She looks at the cafeteria menu to find the possible choices. How many different lunches could she choose?

Today's Lunch Menu

Drink: Milk
Sandwich: Turkey or Tuna
Fruit: Apple or Banana

All the possible choices she could make are called **outcomes**.

You can organize the data on a **tree diagram** to find the outcomes.

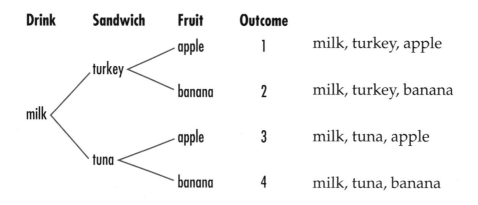

Drink	Sandwich	Fruit	Outcome	
		apple	1	milk, turkey, apple
	turkey	banana	2	milk, turkey, banana
milk		apple	3	milk, tuna, apple
	tuna	banana	4	milk, tuna, banana

Another Example

Suppose you toss a penny and a quarter. A tree diagram can help you see and organize the outcomes.

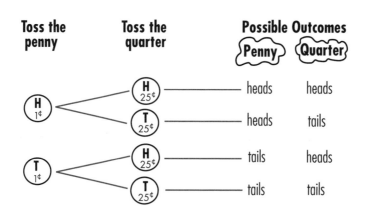

TRY THESE

Use the tree diagram to complete the list of outcomes.

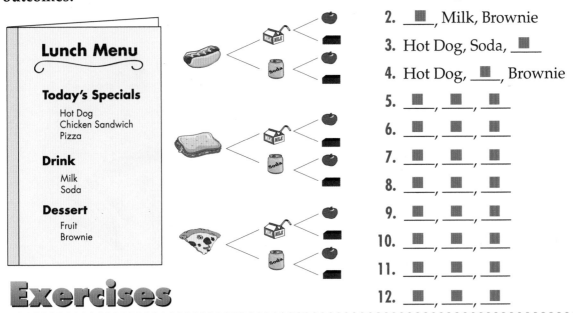

1. Hot Dog, Milk, Fruit
2. ___, Milk, Brownie
3. Hot Dog, Soda, ___
4. Hot Dog, ___, Brownie
5. ___, ___, ___
6. ___, ___, ___
7. ___, ___, ___
8. ___, ___, ___
9. ___, ___, ___
10. ___, ___, ___
11. ___, ___, ___
12. ___, ___, ___

Exercises

Make a tree diagram and list the outcomes.

1. The ice cream shop has 3 kinds of cones: cake, sugar, and wafer. There are 3 flavors of ice cream: vanilla, strawberry, and chocolate. How many different kinds of one-scoop ice cream cones can be made?

2. Sarah has blue, tan, and white slacks. She has brown, red, and yellow shirts. How many different outfits are possible? Name them.

PROBLEM SOLVING

3. The cookie store is having a sale on six kinds of cookies. You can get each kind with raisins, with nuts, with both raisins and nuts, or plain. How many choices do you have?

4. Outcomes are **equally likely** if they have the same probability of occurring. How many equally likely outcomes are there when you spin this spinner?

10.11 Problem-Solving Strategy: Acting It Out

Objective: to solve problems by acting out possible outcomes

Amanda cannot decide which vegetable she wants for supper. Her dad tells her to use a spinner to choose. If she uses the spinner shown at the right, will Amanda ever have the same vegetable more than one night in a row?

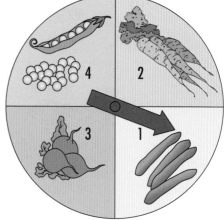

Sometimes you can solve a problem by acting out the situation.

1. READ
You need to find out whether Amanda will have the same vegetable several nights in a row. You know the choices are green beans, carrots, beets, or peas. You know she will choose the vegetables that she spins.

2. PLAN
You need to find out if it is possible to spin the same vegetable several times in a row. You can act out the situation. Spin a spinner with four sections numbered 1, 2, 3, and 4. Record the results.

3. SOLVE
Spin a spinner 48 times and record the results on a tally sheet. Whenever you spin the same number two or more times in a row, put a mark by it.

Vegetable	Tally	Total Number Spun
Green beans	||	
Carrots		
Beets			
Peas			

4. CHECK
It is possible for Amanda to have the same vegetable for several nights in a row.

Each spin is called an **independent event** because one event does not affect the outcome of the other. If you spin a 4, the next spin is equally likely to be a 4.

TRY THESE

Solve by using the act-it-out strategy.

1. At the first stop, some people got on an empty bus. At the second stop, 10 people got on, and 8 people got off. At the next stop, 7 people got on, and 9 people got off. There were 11 people left on the bus. How many people got on at the first stop?

2. Bill and Sue play a game. At the end of the game, the loser gives the winner one chip. Bill has won 3 games, and Sue has 3 more chips than she had at the start. How many games have they played?

Solve

Use any strategy to solve.

1. If you look at a clock in the mirror, it shows 7:10 as the time. What is the correct time?

2. Herb, Ann, and Sandy cross from one bluff to another in a basket on a rope. The basket can carry 160 pounds. Herb weighs 100 pounds; Ann weighs 70 pounds; and Sandy weighs 85 pounds. How can all three cross safely? *Hint:* Each time the basket crosses to a bluff, someone must bring it back so another person can cross.

CONSTRUCTED RESPONSE

3. Hillary and Rick go to the pet store to buy a kitten. There is a cage of 16 kittens. Half are black; $\frac{1}{4}$ are white; and $\frac{1}{4}$ are orange-striped. They are all so cute that Hillary cannot decide which one to choose. She closes her eyes and points to one. Which color kitten did Hillary probably point to? Explain your answer.

TEST PREP

4. Max buys a box of cookies. There are 76 cookies total in the box. Of them, 18 are broken. What are the chances that Max will pick out a broken cookie?

   a. $\frac{76}{18}$ b. $\frac{1}{2}$ c. $\frac{9}{38}$ d. none of the above

LANGUAGE and CONCEPTS

Choose the correct term or number to complete each sentence.

1. In a fraction, the number that represents the part you are talking about is called the _____.

2. In a fraction, the number that represents the equal-sized parts of the whole is called the _____.

3. In the fraction $\frac{7}{15}$, _____ is the total number of objects in the set.

4. A(n) _____ has a whole number part and a fraction part.

5. $\frac{12}{9}$ is a(n) _____.

6. A(n) _____ diagram helps to find outcomes.

15
7
1
3
divide
tree
improper fraction
mixed number
multiply
numerator
denominator

SKILLS and PROBLEM SOLVING

Write a fraction for each colored part. (Section 10.1)

7.

8.

9.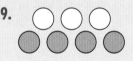

Make a model and find each of the following. (Section 10.2)

10. $\frac{1}{3}$ of 6

11. $\frac{1}{9}$ of 18

12. $\frac{2}{3}$ of 6

13. $\frac{5}{8}$ of 16

14. $\frac{1}{4}$ of 16

15. $\frac{3}{4}$ of 8

Choose two equivalent fractions for the colored part. (Section 10.3)

16.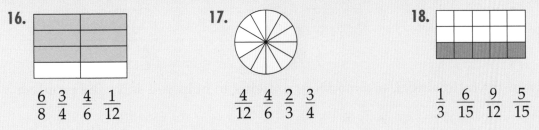

$\frac{6}{8}$ $\frac{3}{4}$ $\frac{4}{6}$ $\frac{1}{12}$

17.

$\frac{4}{12}$ $\frac{4}{6}$ $\frac{2}{3}$ $\frac{3}{4}$

18.

$\frac{1}{3}$ $\frac{6}{15}$ $\frac{9}{12}$ $\frac{5}{15}$

Write a fraction to represent each colored model. Is the fraction in simplest form? Write *yes* or *no*. (Section 10.4)

19.

20.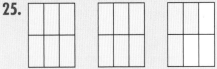

21.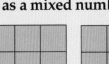

Compare using <, >, or =. (Section 10.5)

22. $\frac{7}{8} \bullet \frac{5}{8}$

23. $\frac{1}{3} \bullet \frac{2}{9}$

24. $\frac{3}{4} \bullet \frac{6}{8}$

Write each colored part as an improper fraction and as a mixed number. (Section 10.6)

25.

26.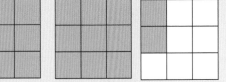

Solve. (Sections 10.7–10.11)

27. Liz has a fruit salad for a snack. There are 6 berries, 2 chunks of pineapple, 8 slices of banana, and 4 orange slivers. What is the probability of Liz eating a berry?

28. James ate $\frac{3}{4}$ of his cookies. If he ate 6 cookies, how many cookies did he have to start?

29. Make a reasonable prediction: If you flip a penny 30 times, how many times would you probably get tails? Act it out to see whether your prediction is true.

30. Joan has red, green, blue, and yellow socks. She has black, brown, and white shoes. Make a tree diagram to show all of the possible outcomes.

Write a fraction for each colored part.

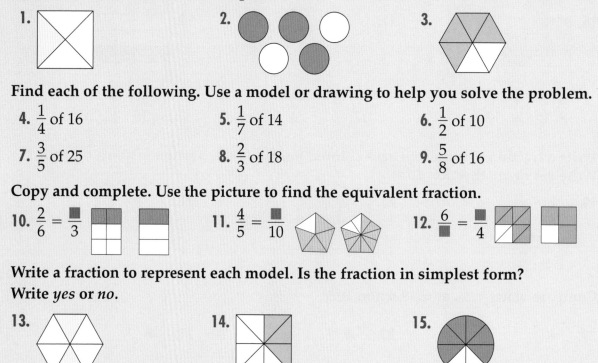

1.

2.

3.

Find each of the following. Use a model or drawing to help you solve the problem.

4. $\frac{1}{4}$ of 16

5. $\frac{1}{7}$ of 14

6. $\frac{1}{2}$ of 10

7. $\frac{3}{5}$ of 25

8. $\frac{2}{3}$ of 18

9. $\frac{5}{8}$ of 16

Copy and complete. Use the picture to find the equivalent fraction.

10. $\frac{2}{6} = \frac{\blacksquare}{3}$

11. $\frac{4}{5} = \frac{\blacksquare}{10}$

12. $\frac{6}{\blacksquare} = \frac{\blacksquare}{4}$

Write a fraction to represent each model. Is the fraction in simplest form? Write *yes* or *no*.

13.

14.

15.

Compare using <, >, or =. Use models or drawings to help you.

16. $\frac{4}{6} \bullet \frac{2}{6}$

17. $\frac{1}{2} \bullet \frac{1}{3}$

18. $\frac{4}{9} \bullet \frac{2}{3}$

Write each improper fraction as a whole number or a mixed number.

19. $\frac{12}{12}$

20. $\frac{9}{5}$

21. $\frac{11}{4}$

22. $\frac{20}{10}$

Solve.

23. Imagine you are rolling a number cube. How many times would you predict that you would roll a 4 if you roll it 30 times?

24. For dinner at the restaurant, you could have chicken, spaghetti, a burger, or a hot dog. You could have mashed potatoes or French fries. You could have applesauce, green beans, or corn for a side dish. Make a tree diagram to show all the possible outcomes of menu choices.

Exploring Proportions

Draw and color the patterns at the right using graph paper and colored pencils.

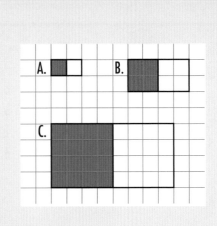

Investigate to find the following.

1. What is the area of figure **A**?

2. How many units in figure **A** are red? Write this as a fraction.

3. What is the area of figure **B**?

4. Figure **B** is an enlargement of figure **A**. How would you show the red part of figure **A** in figure **B**? How many units in **B** are red? Write this as a fraction.

5. Figure **C** is an enlargement of figure **A**. How would you show the red part of figure **A** in figure **C**? How many units in **C** are red? Write this as a fraction.

Notice

A. Figure **B** is in proportion to figure **A** because $\frac{4}{8} = \frac{1}{2}$.

B. Figure **C** is in proportion to figure **A** because $\frac{16}{32} = \frac{1}{2}$.

Decide if the pair of figures makes a proportion. Explain why or why not.

6.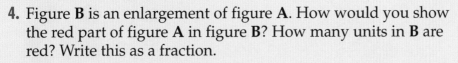

7.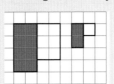

8.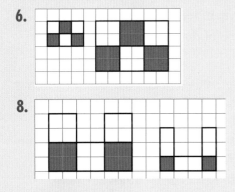

9.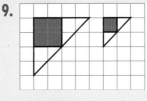

10. **Challenge**
    Copy this pattern. Make an enlarged proportional pattern and a reduced pattern.

11. **Project**
    Use graph paper to make a pattern similar to the figures in problems 6–9. Make an enlarged or reduced copy of each.

CUMULATIVE TEST

1. Which fraction is an equivalent fraction to $\frac{9}{15}$?
   a. $\frac{1}{4}$
   b. $1\frac{6}{15}$
   c. $\frac{3}{5}$
   d. none of the above

2. What is 2,533 rounded to the nearest thousand?
   a. 2,000
   b. 3,000
   c. 4,000
   d. none of the above

3.   2,087
     $\times$ 9
   a. 18,756
   b. 18,783
   c. 81,723
   d. none of the above

4. 8)885
   a. 11
   b. 11 R5
   c. 111 R5
   d. none of the above

5. Which fraction is less than $\frac{2}{3}$?
   a. $\frac{3}{2}$
   b. $\frac{1}{2}$
   c. $\frac{4}{6}$
   d. none of the above

6. What part of the set is *not* yellow?
   a. $\frac{3}{7}$
   b. $\frac{4}{7}$
   c. $\frac{4}{3}$
   d. none of the above

7. Which two figures are polygons?
   a. W, X
   b. W, Y
   c. Y, Z
   d. X, Z

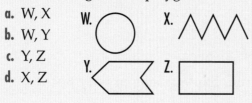

8. There were 42 members in a swim club. Three-sevenths of the members went to a swim meet. How many members went to the meet?
   a. 6 members
   b. 18 members
   c. 24 members
   d. none of the above

9. Tony sold 110 tickets on Wednesday, 112 tickets on Thursday, and 84 tickets on Friday. What was Tony's average number of tickets sold for the 3 days?
   a. 82
   b. 98
   c. 100
   d. 102

10. Ashley collected stickers. Brent collected one-third as many as Ashley. Which statement has the same information?
   a. Ashley has one-third as many stickers as Brent.
   b. Ashley has three times as many stickers as Brent.
   c. Brent has three times as many stickers as Ashley.
   d. Brent has as many stickers as Ashley.

Adding and Subtracting Fractions

Rebecca Hannigan
Maryland

11.1 Adding and Subtracting Fractions with Like Denominators

Objective: to add and subtract fractions with like denominators

Robin baked pies for a party. She cut each pie into eight equal pieces. After the party, she had $\frac{3}{8}$ of the apple pie left and $\frac{4}{8}$ of the cherry pie left. How much of a pie was left?

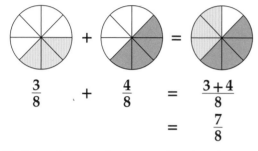 

Robin had $\frac{7}{8}$ of a pie left.

$$\frac{3}{8} + \frac{4}{8} = \frac{3+4}{8}$$
$$= \frac{7}{8}$$

Robin also made brownies for dessert for the party. The brownies were cut into six equal-sized parts $\left(\frac{6}{6}\right)$. Tom ate $\frac{2}{6}$. How much of the brownies was left?

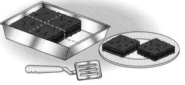

You can use the number line to show subtraction of fractions.

Subtract $\frac{2}{6}$ from $\frac{6}{6}$.

$$\frac{6}{6} - \frac{2}{6} = \frac{6-2}{6} = \frac{4}{6}$$

TRY THESE

Copy and complete.

1.

$$\frac{1}{4} + \frac{2}{4} = \frac{1+2}{4} = \frac{\blacksquare}{\blacksquare}$$

2.

$$\frac{2}{6} + \frac{2}{6} = \frac{\blacksquare}{6}$$

3.

$$\frac{3}{9} + \frac{3}{9} = \frac{\blacksquare}{9}$$

4.

$$\frac{3}{4} - \frac{2}{4} = \frac{3-2}{4} = \frac{\blacksquare}{\blacksquare}$$

5.

$$\frac{5}{8} - \frac{2}{8} = \frac{5-2}{8} = \frac{\blacksquare}{\blacksquare}$$

6.

$$\frac{4}{6} - \frac{3}{6} = \frac{4-3}{6} = \frac{\blacksquare}{\blacksquare}$$

Exercises

Add or subtract.

1. $\frac{3}{8} + \frac{2}{8}$

2. $\frac{1}{6} + \frac{1}{6}$

3. $\frac{3}{10} + \frac{2}{10}$

4. $\frac{3}{7} + \frac{2}{7}$

5. $\frac{6}{6} - \frac{4}{6}$

6. $\frac{13}{16} - \frac{1}{16}$

7. $\frac{7}{15} - \frac{3}{15}$

8. $\frac{7}{12} - \frac{4}{12}$

9. $\begin{array}{r} \frac{3}{12} \\ + \frac{7}{12} \\ \hline \end{array}$

10. $\begin{array}{r} \frac{13}{16} \\ - \frac{7}{16} \\ \hline \end{array}$

11. $\begin{array}{r} \frac{2}{10} \\ + \frac{6}{10} \\ \hline \end{array}$

12. $\begin{array}{r} \frac{4}{4} \\ - \frac{2}{4} \\ \hline \end{array}$

13. $\begin{array}{r} \frac{7}{10} \\ - \frac{3}{10} \\ \hline \end{array}$

PROBLEM SOLVING

Use the shopping list to solve.

14. How many total pounds of lunch meat were bought?

15. How much more Swiss cheese than Cheddar was bought?

16. If Jack uses $\frac{1}{4}$ pound of bologna to make sandwiches, how much bologna will be left?

17. What is the total weight of cheese bought?

★ 18. Was more cheese or lunch meat bought? How much more? Draw diagrams to show the total amount of each, and then compare the diagrams.

shopping List

Cheese
$\frac{2}{3}$ pound Swiss
$\frac{1}{3}$ pound Cheddar
$\frac{2}{3}$ pound Colby

Lunch Meat
$\frac{1}{4}$ pound Salami
$\frac{3}{4}$ pound Ham
$\frac{2}{4}$ pound Bologna

MIND BUILDER

Rounding Fractions

You can see on the number line that fractions are closer to $0, \frac{1}{2}$, or 1.

$\frac{1}{12}$ is about 0. $\frac{2}{6}$ is about $\frac{1}{2}$. $\frac{5}{6}$ is about 1.

Use the number line above to decide whether each of these fractions is about 0, about $\frac{1}{2}$, or about 1.

1. $\frac{1}{6}$

2. $\frac{3}{12}$

3. $\frac{4}{6}$

4. $\frac{2}{3}$

5. $\frac{3}{4}$

6. $\frac{5}{12}$

11.2 Adding and Subtracting Mixed Numbers with Like Denominators

Objective: to add and subtract mixed numbers with like denominators

One brownie recipe calls for $2\frac{1}{4}$ cups of flour.
Another brownie recipe calls for $1\frac{1}{4}$ cups of flour.
How much flour is needed for both recipes?

Add $2\frac{1}{4} + 1\frac{1}{4}$.

Step 1	Step 2	Step 3
Add the numerators. Then write the sum over the denominator.	Add the ones.	Write the fraction in simplest form.
$\begin{aligned} 2\frac{1}{4} \\ + 1\frac{1}{4} \\ \hline \frac{2}{4} \end{aligned}$	$\begin{aligned} 2\frac{1}{4} \\ + 1\frac{1}{4} \\ \hline 3\frac{2}{4} \end{aligned}$	$\begin{aligned} 2\frac{1}{4} \\ + 1\frac{1}{4} \\ \hline 3\frac{2}{4} \text{ or } 3\frac{1}{2} \end{aligned}$ $\frac{2 \div 2}{4 \div 2} = \frac{1}{2}$

The two recipes need $3\frac{1}{2}$ cups of flour.

More Examples

A. $2\frac{1}{6} + 3\frac{4}{6} = 5\frac{5}{6}$

B. $\begin{aligned} 4\frac{7}{9} \\ - 2\frac{2}{9} \\ \hline 2\frac{5}{9} \end{aligned}$

C. $\begin{aligned} 6\frac{4}{8} \\ - 2\frac{2}{8} \\ \hline 4\frac{2}{8} \text{ or } 4\frac{1}{4} \end{aligned}$

D. $\begin{aligned} 3\frac{1}{4} \\ + 2\frac{3}{4} \\ \hline 5\frac{4}{4} \text{ or } 6 \end{aligned}$

TRY THESE

Add or subtract. Write the answers in simplest form.

1. $\begin{aligned} 3\frac{1}{3} \\ + 1\frac{1}{3} \end{aligned}$

2. $\begin{aligned} 5\frac{7}{8} \\ - 2\frac{2}{8} \end{aligned}$

3. $\begin{aligned} 4\frac{2}{5} \\ + 2\frac{1}{5} \end{aligned}$

4. $\begin{aligned} 7\frac{5}{9} \\ - 3\frac{3}{9} \end{aligned}$

5. $\begin{aligned} 2\frac{3}{7} \\ + 3\frac{2}{7} \end{aligned}$

Exercises

Add or subtract. Write the answers in simplest form.

1. $5\frac{1}{7}$
   $+ 3\frac{3}{7}$

2. $7\frac{7}{8}$
   $- 2\frac{1}{8}$

3. $4\frac{1}{3}$
   $+ 3\frac{2}{3}$

4. $6\frac{8}{9}$
   $- 4\frac{3}{9}$

5. $6\frac{5}{7}$
   $+ 3\frac{2}{7}$

6. $7\frac{6}{10}$
   $- 4\frac{3}{10}$

7. $2\frac{1}{3}$
   $+ 6\frac{2}{3}$

8. $9\frac{5}{8}$
   $- 4\frac{1}{8}$

9. $4\frac{1}{6}$
   $+ 3\frac{2}{6}$

10. $7\frac{3}{4}$
    $- 2\frac{1}{4}$

11. $7\frac{3}{5}$
    $+ 1\frac{1}{5}$

12. $6\frac{5}{6}$
    $- 1\frac{4}{6}$

13. $4\frac{6}{9}$
    $+ 5\frac{2}{9}$

14. $9\frac{7}{8}$
    $- 4\frac{5}{8}$

15. $9\frac{5}{6}$
    $- 2\frac{1}{6}$

16. $6\frac{2}{4} + 2\frac{1}{4}$

17. $4\frac{4}{5} - 2\frac{1}{5}$

18. $5\frac{4}{6} + 3\frac{2}{6}$

19. $15\frac{7}{10} - 6\frac{2}{10}$

20. $12\frac{3}{8} + 4\frac{4}{8}$

★ 21. $5\frac{1}{6} - 3\frac{1}{6}$

★ 22. $4\frac{3}{5} + 9$

★ 23. $7\frac{2}{4} - 3$

PROBLEM SOLVING

24. The members of the Ecology Club are helping to plant trees. They plant $2\frac{5}{12}$ boxes of seedlings on the first day and $2\frac{1}{12}$ boxes on the next day. How many total boxes did they plant?

25. In the morning, Sally uses $2\frac{4}{8}$ bags of fertilizer. She uses $1\frac{3}{8}$ bags that afternoon. How much fertilizer does she use in all?

26. Use the information from **Problem Solving 25**. How much more fertilizer did Sally use in the morning than in the afternoon?

★ 27. On Friday it snowed $2\frac{1}{10}$ inches. On Saturday and Sunday, it snowed $5\frac{3}{10}$ each day. How much did it snow all 3 days?

★ 28. Tricia ate $\frac{1}{5}$ of the pumpkin pie on Sunday, $\frac{1}{5}$ on Monday, and $\frac{1}{5}$ on Tuesday. How much of the pie was left?

11.2 Adding and Subtracting Mixed Numbers with Like Denominators **287**

11.3 Problem-Solving Application: Two-Step Problems

Objective: to solve problems with two steps

The Dobels used $\frac{2}{8}$ of a loaf of bread for breakfast and $\frac{4}{8}$ of a loaf for lunch. How much of the loaf was left after lunch?

Solve the problem using the following plan.

1. READ You need to find how much of the loaf was left after lunch. You know what part of the loaf of bread was used at each meal.

2. PLAN You cannot find how much was left until you find the total amount used. The problem takes two steps to solve. First add. Then subtract to find what is left. Write an equation. $n = \frac{8}{8} - \left(\frac{2}{8} + \frac{4}{8} \right)$

3. SOLVE First add $\frac{2}{8}$ and $\frac{4}{8}$.

$\frac{2}{8} + \frac{4}{8} = \frac{6}{8}$ The Dobels used $\frac{6}{8}$ of a loaf of bread for two meals.

> ▶ Work the operation inside the parentheses first.

Then subtract $\frac{6}{8}$ from a whole loaf $\left(\frac{8}{8} \right)$.

$\frac{8}{8} - \frac{6}{8} = \frac{2}{8}$ $\frac{2}{8}$ of the loaf was left after lunch.

4. CHECK Add what was left, $\frac{2}{8}$, to what was used at breakfast, $\frac{2}{8}$, and what was used at lunch, $\frac{4}{8}$.

$\frac{2}{8} + \frac{2}{8} + \frac{4}{8} = \frac{8}{8}$ or 1 whole loaf

TRY THESE

Solve.

1. Su-Ming used $\frac{2}{6}$ of a bag of flour for baking a cake and $\frac{1}{6}$ of a bag of flour for cookies. How much of a full bag of flour was left after baking?

2. Tom used $\frac{6}{10}$ lb of cherries for making a dessert and $\frac{3}{10}$ lb for a mixed fruit salad. What part of a pound was left from a 1-lb box of cherries?

1. How much cheese is shown in all?

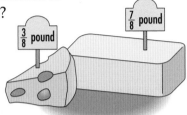

2. Trudi made a fruit salad by combining these strawberries and blueberries. Did she make more than a pound of fruit salad?

3. Maria ate $\frac{2}{8}$ of the pizza, and Tony ate $\frac{3}{8}$ of the pizza. How much of the pizza was left?

4. Jacob bought 1 cup of butter and used $\frac{1}{8}$ of a cup of it. How much butter did he have left?

5. It snowed 11 inches in December, 15 inches in January, and 23 inches in February. What was the total snowfall for these 3 months?

6. Brian practiced playing the flute for 35 minutes on Monday, 45 minutes on Tuesday, and 40 minutes on Thursday. What is the average number of minutes he practiced each day?

7. A raccoon's hind paw is about $3\frac{6}{8}$ inches long. A mink's hind paw is about $1\frac{2}{8}$ inches long. About how much longer is the raccoon's paw than the mink's paw?

CONSTRUCTED RESPONSE

8. At a bake sale, Mrs. Hawkins sold $\frac{4}{6}$ of one pie, $\frac{3}{6}$ of another pie, and $\frac{2}{6}$ of a third pie. How much pie did Mrs. Hawkins sell in all? Explain how you found your answer.

MIXED REVIEW

Add or subtract. Write the answers in simplest form.

9. $\frac{1}{3} + \frac{1}{3}$

10. $\frac{5}{12} + \frac{3}{12}$

11. $4\frac{4}{9} + 5\frac{2}{9}$

12. $3\frac{5}{7} + 2\frac{2}{7}$

13. $\frac{6}{9} - \frac{3}{9}$

14. $\frac{4}{5} - \frac{2}{5}$

15. $15\frac{9}{10} - 8\frac{5}{10}$

16. $9\frac{15}{16} - 6\frac{7}{16}$

Solve.

17. Mr. Jones spent $2\frac{2}{4}$ hours mowing the lawn and another $1\frac{1}{4}$ hours weeding. How much time did he spend mowing and weeding?

11.4 Exploring Unlike Denominators

Objective: to compare fractions with unlike denominators

Hadley has $\frac{2}{3}$ of her sandwich left. Tyler has $\frac{1}{2}$ of his sandwich left. Who has more left?

When you compare fractions, it is best to make the fractions have the same denominators.

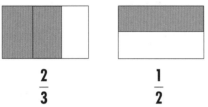

$$\frac{2}{3} \qquad \frac{1}{2}$$

Look at the denominators. What are the multiples of each?

    3: **3, 6, 9, 12, 15, 18**

    2: **2, 4, 6, 8, 10, 12, 14, 16, 18**

Find a multiple that 3 and 2 share. Use the smallest one, which is 6.

$$\frac{2}{3} = \frac{\blacksquare}{6} \qquad 3 \times \blacksquare = 6$$

If you change the denominators to 6, you need to change the numerators to make an equivalent fraction.

$$3 \times 2 = 6$$

$$\frac{2}{3} = \frac{2}{3} \times \frac{2}{2} = \frac{4}{6}$$ You need to multiply 2×2 to get a new numerator and another equivalent fraction.

Now find an equivalent fraction to $\frac{1}{2}$ with 6 in the denominator. To do that, multiply both the top and the bottom by 3. When you do that, you are multiplying by $\frac{3}{3}$. It is the same as multiplying by 1.

$$\frac{1}{2} = \frac{\blacksquare}{6} = \frac{1}{2} \times \frac{3}{3} = \frac{1 \times 3}{2 \times 3} = \frac{3}{6}$$

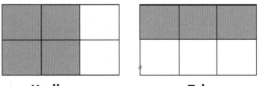

Hadley   **Tyler**

Now compare the fractions $\frac{4}{6}$ and $\frac{3}{6}$.
Hadley has $\frac{4}{6}$ of her sandwich left.
Tyler has $\frac{3}{6}$ of his sandwich left.
Hadley has more of her sandwich left.

TRY THESE

Change these fractions to fractions with like denominators.
Circle the smaller fraction. Be sure to show your work.

1. $\frac{3}{4}$ and $\frac{5}{8}$ 2. $\frac{1}{4}$ and $\frac{2}{3}$ 3. $\frac{1}{2}$ and $\frac{1}{5}$ 4. $\frac{1}{3}$ and $\frac{4}{6}$

Exercises

Change these fractions to fractions with like denominators.
Circle the greater fraction.

1. $\frac{3}{6}$ and $\frac{5}{8}$ 2. $\frac{3}{4}$ and $\frac{1}{5}$ 3. $\frac{2}{9}$ and $\frac{1}{2}$ 4. $\frac{1}{6}$ and $\frac{2}{3}$

5. $\frac{11}{14}$ and $\frac{3}{7}$ 6. $\frac{1}{2}$ and $\frac{4}{7}$ 7. $\frac{7}{8}$ and $\frac{4}{5}$ 8. $\frac{10}{10}$ and $\frac{5}{5}$

PROBLEM SOLVING

9. Roy painted $\frac{1}{4}$ of the dining room. Jennifer painted $\frac{3}{8}$ of the living room. If the rooms are the same size, who painted more?

10. Tim has mowed $\frac{4}{5}$ of the lawn. Faith has weeded $\frac{1}{2}$ of the garden. Who has finished more yard work?

11. There are 5 balloons. Dawn filled $\frac{2}{5}$ of the balloons with water and filled the remaining balloons with air. How many balloons are filled with water?

CONSTRUCTED RESPONSE

12. Look at the answer for **Problem Solving 11**. If Dawn doubled the amount of balloons for a party, how would the results change? Explain how you could figure out the answer without changing the denominator.

11.4 Exploring Unlike Denominators   291

An Event That Will Occur

Theresa is having a party. She wants to make sure that 2 people there have birthdays in the same month. What is the least number of people that Theresa needs to invite to her party?

Extension

Change the problem to read, "on the same day of the week," instead of, "in the same month." Then solve.

CUMULATIVE REVIEW

Copy and complete.

1. 9 + ■ = 18

2. 8 + 6 = 6 + ■

3. ■ + 7 = 7

Find the sum or difference.

4.	7,268
	− 1,672

5.	1,377
	+ 3,726

6.	555
	38
	+ 294

7.	7,600
	− 3,145

8.	271
	+ 438

9. $3.78 + $4.99

10. $34.95 + $2.76

11. $24.27 − $15.32

Write in expanded form.

12. 237

13. 3,582

14. 28,074

15. 691

Multiply.

16. 9 × 9

17. 7 × 6

18. 8 × 7

19. 6 × 9

20. 5 × 7

21.	67
	× 28

22.	$4.54
	× 12

23.	2,457
	× 8

24.	5,709
	× 34

25.	$12.95
	× 6

Solve.

26. What was the midday temperature on Wednesday?

27. Was it warmer on Monday or Thursday?

28. On what day was there the lowest temperature?

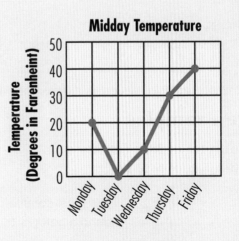

Midday Temperature

Temperature (Degrees in Farenheint)

Days of the Week

11.5 Adding Fractions with Unlike Denominators

Objective: to add fractions with unlike denominators

Caroline bought $\frac{2}{3}$ of a pound of green apples and $\frac{1}{4}$ of a pound of red apples. How many pounds of apples did she buy in all?

Add $\frac{2}{3}$ and $\frac{1}{4}$.

Because $\frac{2}{3}$ and $\frac{1}{4}$ have unlike denominators, you need to rename them using equivalent fractions that have the same denominator. Use the least common denominator.

The **least common denominator** is the least common multiple of two or more denominators.

Step 1	Step 2	Step 3
List the multiples to find the least common denominator.	Write equivalent fractions.	Add.
$\frac{2}{3}$ multiples of 3: 3, 6, 9, 12, ... $+\frac{1}{4}$ multiples of 4: 4, 8, 12, 16, ... The least common denominator is 12.	$\dfrac{2 \times 4}{3 \times 4} = \dfrac{8}{12}$ $+\dfrac{1 \times 3}{4 \times 3} = \dfrac{3}{12}$	$\dfrac{8}{12}$ $+\dfrac{3}{12}$ $\dfrac{11}{12}$

Caroline bought $\frac{11}{12}$ (eleven-twelfths) of a pound of apples.

More Examples

A. multiples of 3: 3, 6, 9, 12
multiples of 6: 6, 12, 18

$\begin{aligned}\frac{2}{3} &= \frac{4}{6}\\ +\frac{1}{6} &= \frac{1}{6}\\ \hline &\frac{5}{6}\end{aligned}$ The least common denominator is 6.

B. multiples of 5: 5, 10, 15
multiples of 10: 10, 20, 30

$\begin{aligned}\frac{2}{5} &= \frac{4}{10}\\ +\frac{2}{10} &= \frac{2}{10}\\ \hline \frac{6}{10} &= \frac{3}{5}\end{aligned}$ The least common denominator is 10.

TRY THESE

Find the least common denominator.

1. $\dfrac{1}{4}, \dfrac{1}{2}$ 2. $\dfrac{1}{5}, \dfrac{1}{3}$ 3. $\dfrac{1}{6}, \dfrac{1}{4}$ 4. $\dfrac{1}{3}, \dfrac{1}{6}$

Exercises

Add. Write the sum in simplest form.

1. $\dfrac{1}{4} = \dfrac{3}{12}$
   $+\dfrac{2}{3} = \dfrac{8}{12}$

2. $\dfrac{1}{2}$
   $+\dfrac{1}{3}$

3. $\dfrac{2}{5}$
   $+\dfrac{3}{10}$

4. $\dfrac{3}{8}$
   $+\dfrac{2}{4}$

5. $\dfrac{1}{4}$
   $+\dfrac{2}{7}$

6. $\dfrac{1}{2}$
   $+\dfrac{1}{6}$

7. $\dfrac{3}{4}$
   $+\dfrac{1}{12}$

8. $\dfrac{4}{9}$
   $+\dfrac{1}{2}$

9. $\dfrac{1}{4}$
   $+\dfrac{2}{8}$

10. $\dfrac{2}{5}$
    $+\dfrac{2}{7}$

11. $\dfrac{7}{12} + \dfrac{1}{6}$ 12. $\dfrac{1}{3} + \dfrac{5}{6}$ 13. $\dfrac{3}{5} + \dfrac{7}{10}$ ★14. $\dfrac{1}{2} + \dfrac{1}{4} + \dfrac{1}{8}$

★15. Find the sum of $\dfrac{2}{3} + \dfrac{1}{2} + \dfrac{1}{6}$ in simplest form.

PROBLEM SOLVING

16. Collin bought $\dfrac{5}{8}$ of a yard of red ribbon and $\dfrac{1}{2}$ of a yard of green ribbon. How much ribbon did he buy in all?

17. Elizabeth used $\dfrac{5}{6}$ of a bag of flour for one batch of cookies and $\dfrac{3}{9}$ of a bag on the second batch. How much total flour did she use?

TEST PREP

18. On Monday, Lori walked $\dfrac{2}{10}$ of a mile. On Tuesday, she walked $\dfrac{2}{3}$ of a mile. On Wednesday, she walked $\dfrac{1}{2}$ of a mile. By Thursday, how far had she already walked that week?

   a. 1 mile   b. $1\dfrac{1}{2}$ miles   c. $\dfrac{17}{15}$ miles   d. $1\dfrac{11}{30}$ miles

11.6 Subtracting Fractions with Unlike Denominators

Objective: to subtract fractions with unlike denominators

Frances had $\frac{1}{2}$ of a box of pancake mix. She used $\frac{1}{4}$ of the mix to make pancakes. What part of the mix was left?

Subtract $\frac{1}{4}$ from $\frac{1}{2}$.

Because $\frac{1}{4}$ and $\frac{1}{2}$ have unlike denominators, you have to rename them using equivalent fractions with a least common denominator.

Step 1	Step 2	Step 3
List the multiples to find the least common denominator.	Write equivalent fractions.	Subtract.
$\frac{1}{2}$ multiples of 2: 2, 4, 6, 8, . . . $-\frac{1}{4}$ multiples of 4: 4, 8, 12, 16, . . . The least common denominator is 4.	$\begin{array}{c} \frac{1}{2} = \frac{2}{4} \\[6pt] -\frac{1}{4} = \frac{1}{4} \end{array}$	$\begin{array}{c} \frac{2}{4} \\[6pt] -\frac{1}{4} \\ \hline \frac{1}{4} \end{array}$

There was $\frac{1}{4}$ of the box of pancake mix left.

More Examples

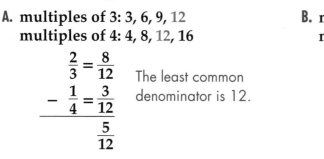

A. multiples of 3: 3, 6, 9, 12
multiples of 4: 4, 8, 12, 16

$$\begin{array}{c} \frac{2}{3} = \frac{8}{12} \\[6pt] -\frac{1}{4} = \frac{3}{12} \\ \hline \frac{5}{12} \end{array}$$

The least common denominator is 12.

B. multiples of 4: 4, 8, 12, 16
multiples of 8: 8, 16, 24

$$\begin{array}{c} \frac{3}{4} = \frac{6}{8} \\[6pt] -\frac{1}{8} = \frac{1}{8} \\ \hline \frac{5}{8} \end{array}$$

The least common denominator is 8.

TRY THESE

Find the least common denominator.

1. $\dfrac{4}{5}, \dfrac{1}{2}$ 2. $\dfrac{1}{3}, \dfrac{1}{2}$ 3. $\dfrac{5}{6}, \dfrac{2}{3}$ 4. $\dfrac{4}{5}, \dfrac{1}{3}$ 5. $\dfrac{2}{5}, \dfrac{3}{4}$

Exercises

Subtract. Write the difference in simplest form.

1. $\begin{aligned} \dfrac{4}{5} &= \dfrac{8}{10} \\ -\dfrac{1}{2} &= \dfrac{5}{10} \end{aligned}$

2. $\begin{aligned} \dfrac{5}{8} &= \dfrac{5}{8} \\ -\dfrac{1}{4} &= \dfrac{2}{8} \end{aligned}$

3. $\begin{aligned} \dfrac{1}{3} \\ -\dfrac{2}{9} \end{aligned}$

4. $\begin{aligned} \dfrac{3}{8} \\ -\dfrac{1}{4} \end{aligned}$

5. $\begin{aligned} \dfrac{5}{6} \\ -\dfrac{1}{2} \end{aligned}$

6. $\begin{aligned} \dfrac{1}{3} \\ -\dfrac{1}{5} \end{aligned}$

7. $\begin{aligned} \dfrac{3}{4} \\ -\dfrac{1}{2} \end{aligned}$

8. $\begin{aligned} \dfrac{2}{3} \\ -\dfrac{1}{6} \end{aligned}$

9. $\begin{aligned} \dfrac{2}{4} \\ -\dfrac{1}{12} \end{aligned}$

10. $\begin{aligned} \dfrac{4}{5} \\ -\dfrac{2}{4} \end{aligned}$

Complete the following problems.

11. What is the first step in subtracting fractions with unlike denominators?

★12. Find $\dfrac{1}{2}$ plus $\dfrac{2}{4}$ minus $\dfrac{2}{3}$.

PROBLEM SOLVING

13. Andrea walks $\dfrac{5}{10}$ of a mile to school. Harrison walks $\dfrac{2}{5}$ of a mile to school. Who walks farther? How much farther?

14. Frank had $\dfrac{5}{8}$ of a candy bar. He ate $\dfrac{1}{4}$ more. How much does he have left?

MID-CHAPTER REVIEW

Add or subtract. Write the answers in simplest form.

1. $3\dfrac{3}{4} + 2\dfrac{2}{4}$

2. $6\dfrac{3}{7} - 4\dfrac{1}{7}$

3. $\dfrac{1}{4} + \dfrac{4}{5}$

4. $\dfrac{5}{6} - \dfrac{2}{3}$

5. $\dfrac{1}{4} + \dfrac{3}{8}$

6. $\dfrac{4}{5} - \dfrac{3}{4}$

11.7 Adding and Subtracting Mixed Numbers with Unlike Denominators

Objective: to add and subtract mixed numbers with unlike denominators

Mixed numbers with unlike denominators can be added or subtracted.

Imagine you watched one TV show that was $1\frac{1}{4}$ hours long and another show that was $2\frac{1}{2}$ hours long. How long did you spend watching TV? You can figure that out by adding $1\frac{1}{4}$ and $2\frac{1}{2}$.

Step 1	Step 2	Step 3
List the multiples to find the least common denominator.	Write equivalent fractions.	Add.
$1\frac{1}{4}$ 4: 4, 8, 12, ... $+ 2\frac{1}{2}$ 2: 2, 4, 6, ... The least common denominator is 4.	$1\frac{1}{4} = 1\frac{1}{4}$ $+ 2\frac{1}{2} = 2\frac{2}{4}$	$1\frac{1}{4}$ $+ 2\frac{2}{4}$ $3\frac{3}{4}$

You watched $3\frac{3}{4}$ hours of TV.

More Examples

A.

$$2\frac{2}{5} = 2\frac{6}{15}$$
$$+ \ 1\frac{1}{3} = 1\frac{5}{15}$$
$$\overline{\phantom{+ \ 1\frac{1}{3} = } 3\frac{11}{15}}$$

B.

$$4\frac{3}{4} = 4\frac{9}{12}$$
$$- \ 3\frac{1}{6} = 3\frac{2}{12}$$
$$\overline{\phantom{- \ 3\frac{1}{6} = } 1\frac{7}{12}}$$

TRY THESE

Find the sum or difference. Write each answer in simplest form.

1. $2\frac{4}{5} + 2\frac{3}{4}$

2. $6\frac{4}{5} - 2\frac{3}{4}$

3. $3\frac{1}{2} + 4\frac{1}{4}$

4. $4\frac{1}{2} - 3\frac{1}{4}$

5. $4\frac{4}{5} + 3\frac{3}{4}$

6. $4\frac{4}{5} - 3\frac{3}{6}$

7. $1\frac{1}{2} + 8\frac{1}{4}$

8. $8\frac{5}{8} - 4\frac{1}{2}$

Exercises

Find the sum or difference. Write each answer in simplest form.

1. $3\frac{1}{7}$
$+\,6\frac{2}{5}$

2. $7\frac{3}{4}$
$-\,3\frac{1}{8}$

3. $4\frac{1}{3}$
$+\,9\frac{3}{6}$

4. $8\frac{4}{5}$
$-\,6\frac{1}{3}$

5. $2\frac{1}{2}$
$+\,6\frac{1}{4}$

6. $9\frac{2}{3}$
$-\,5\frac{1}{4}$

7. $3\frac{4}{8}$
$+\,5\frac{1}{4}$

8. $8\frac{3}{4}$
$-\,6\frac{3}{10}$

9. $5\frac{1}{6}$
$+\,10\frac{2}{8}$

10. $7\frac{3}{4}$
$-\,2\frac{1}{2}$

PROBLEM SOLVING

11. A ham has been in the oven for $1\frac{1}{3}$ hours. It needs to cook for $2\frac{1}{2}$ hours in all. How much longer does the ham need to cook?

12. Ryan walked $2\frac{1}{5}$ miles yesterday and another $4\frac{1}{3}$ miles today. How far has Ryan walked in 2 days?

white sugar $3\frac{1}{4}$ cups

brown sugar $1\frac{1}{3}$ cups

13. How much sugar is in both of the canisters on the right?

TEST PREP

14. Steve pours 3 glasses of milk that are each $\frac{1}{2}$ full. He also pours one whole glass of milk. How much milk does Steve pour altogether?

   a. 3 glasses of milk   **b.** $2\frac{1}{2}$ glasses of milk   **c.** 2 glasses of milk

15.
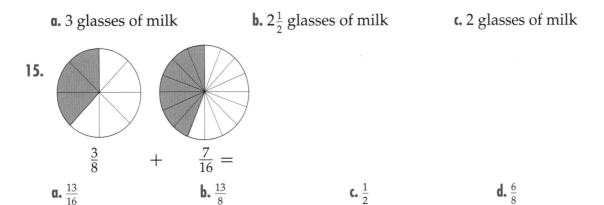

$\frac{3}{8}$ $+$ $\frac{7}{16}$ $=$

a. $\frac{13}{16}$ **b.** $\frac{13}{8}$ **c.** $\frac{1}{2}$ **d.** $\frac{6}{8}$

11.8 Problem-Solving Application: Using Fractions

Objective: to solve problems involving fractions

Jeff is grilling hot dogs on the grill. He used one whole package of beef hot dogs, $\frac{3}{4}$ of the package of turkey hot dogs, and $\frac{3}{8}$ of the package of pork hot dogs. How many packages of hot dogs did Jeff use altogether?

1. READ	Read the problem. You need to find how many total packages of hot dogs Jeff grilled.

2. PLAN	Decide which operation to use to find the total. You add to find the total amount. Write an equation.

$$n = 1 + \frac{3}{4} + \frac{3}{8}$$

3. SOLVE	To add fractions, you need to have a common denominator.

$$n = 1 + \frac{3}{4} + \frac{3}{8} \longrightarrow n = \frac{8}{8} + \frac{6}{8} + \frac{3}{8}$$

Add. $\frac{17}{8}$ Is the fraction in simplest form? No.

$$\frac{17}{8} = 2\frac{1}{8}$$

Jeff used $2\frac{1}{8}$ packages of hot dogs.

4. CHECK	Is your answer reasonable? $2\frac{1}{8}$ is less than 3. You know that Jeff had 3 different packages of hot dogs, so $2\frac{1}{8}$ packages would make sense. You also could draw a picture to check your answer.

TRY THESE
....

Write *true* or *false*.

1. $\frac{4}{5} + \frac{2}{3} = 1\frac{7}{15}$ **2.** $\frac{6}{7} - \frac{3}{5} = 1\frac{3}{5}$ **3.** $\frac{5}{6} + \frac{1}{6} = 2$ **4.** $12\frac{3}{4} - 6\frac{1}{2} = 6\frac{1}{4}$

Solve ....

1. The recipe for chocolate cookies calls for $\frac{1}{2}$ of a cup of white sugar and $\frac{3}{4}$ of a cup of brown sugar. How much sugar will be used altogether?

2. The same chocolate cookie recipe calls for $2\frac{1}{4}$ cups of flour. Sam has two partly filled bags of flour. One has $\frac{2}{3}$ of a cup, and the other one has $1\frac{1}{2}$ cups of flour. Does he have enough flour for the recipe?

CONSTRUCTED RESPONSE

3. Each batch of chocolate cookies makes 24 cookies. If Sam and Will make $2\frac{1}{2}$ batches of cookies, how many cookies will they make? Will they have enough cookies to each have one and share with their teacher and their 23 classmates? Explain.

MIXED REVIEW

Name each shape.

4. 5. 6. 7. 8.

MIND BUILDER

Fraction Squares

Copy. Add across and down. Simplify.

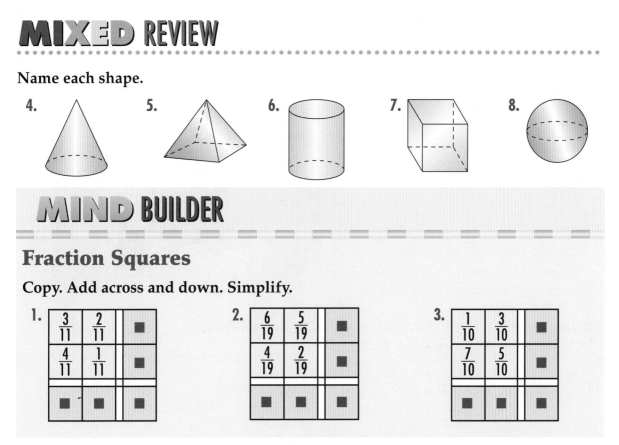

1.

$\frac{3}{11}$	$\frac{2}{11}$	■
$\frac{4}{11}$	$\frac{1}{11}$	■
■	■	■

2.

$\frac{6}{19}$	$\frac{5}{19}$	■
$\frac{4}{19}$	$\frac{2}{19}$	■
■	■	■

3.

$\frac{1}{10}$	$\frac{3}{10}$	■
$\frac{7}{10}$	$\frac{5}{10}$	■
■	■	■

11.8 Problem-Solving Application: Using Fractions 301

LANGUAGE and CONCEPTS

Write the letter of the word that best completes each sentence.

1. When you add or subtract fractions, you need to have a common _____.

2. $\frac{3}{8}$ is _____ to $\frac{6}{16}$.

3. To add $\frac{7}{15}$ and $\frac{6}{7}$, first you need to think of the _____ for both 15 and 7.

a. multiples
b. denominator
c. numerator
d. equivalent

SKILLS and PROBLEM SOLVING

Add or subtract. Write your answers in simplest form. (Sections 11.1–11.2)

4. $\frac{2}{3} + \frac{1}{3}$

5. $11\frac{3}{4} + 4\frac{1}{4}$

6. $\frac{6}{7} - \frac{3}{7}$

7. $8\frac{2}{3} - 4\frac{1}{3}$

8. $14\frac{7}{10} - 9\frac{3}{10}$

Solve. (Section 11.3)

9. Jerry is fifth in line. If everyone in the line turns around, Jerry will be twelfth in line. How many students are standing in line?

10. Danny read $\frac{4}{9}$ of a book before dinner. He read $\frac{2}{9}$ of it after dinner. How much of his book does he still have to read?

Change these fractions to fractions with like denominators. Circle the greater fraction. (Section 11.4)

11. $\frac{2}{5}$ and $\frac{1}{3}$

12. $\frac{4}{7}$ and $\frac{1}{2}$

13. $\frac{1}{12}$ and $\frac{5}{6}$

14. $\frac{1}{10}$ and $\frac{3}{4}$

Add or subtract. Find the common denominators first. Write each answer in simplest form. (Sections 11.5–11.7)

15. $\frac{1}{4} + \frac{5}{6}$

16. $\frac{5}{6} - \frac{1}{9}$

17. $\frac{7}{9} - \frac{2}{6}$

18. $3\frac{3}{4} + 3\frac{1}{6}$

19. $9\frac{1}{4} - 3\frac{1}{6}$

20. $1\frac{1}{4} + 3\frac{3}{5}$

21. $7\frac{3}{14} + 2\frac{1}{7}$

22. $12\frac{4}{5} - 8\frac{1}{2}$

Solve. (Section 11.8)

23. Denise ran $\frac{1}{2}$ of a mile. David ran $\frac{2}{3}$ of a mile. How much did they run altogether?

24. Eric took $\frac{3}{4}$ of a loaf of bread to feed the ducks. He used $\frac{1}{8}$ of the loaf. How much bread was left?

25. Joe needs $\frac{5}{8}$ of a cup of oil for his car. He has put in $\frac{1}{4}$ of a cup of oil. How much more oil does he need?

26. Juan is flying from London to New York. He napped during the first $\frac{1}{8}$ of his trip, and he read during the next $\frac{1}{3}$ of his trip. How much more of his flight does he have left?

CHAPTER 11 TEST

Change these fractions to fractions with like denominators.
Circle the smaller fraction.

1. $\frac{2}{7}$ and $\frac{2}{3}$ 2. $\frac{1}{4}$ and $\frac{1}{6}$ 3. $\frac{3}{4}$ and $\frac{7}{12}$ 4. $\frac{1}{3}$ and $\frac{7}{8}$

Add. Write each sum in simplest form.

5. $\frac{1}{6} + \frac{2}{6}$ 6. $\frac{3}{6} + \frac{1}{2}$ 7. $\frac{3}{8} + \frac{4}{5}$ 8. $\frac{3}{9} + \frac{5}{6}$ 9. $\frac{6}{10} + \frac{2}{5}$

10. $\begin{array}{r} \frac{6}{7} \\ + \frac{1}{7} \\ \hline \end{array}$ 11. $\begin{array}{r} 5\frac{6}{7} \\ + 8\frac{1}{7} \\ \hline \end{array}$ 12. $\begin{array}{r} 5\frac{4}{6} \\ + 8\frac{1}{7} \\ \hline \end{array}$ 13. $\begin{array}{r} 1\frac{5}{8} \\ + 2\frac{2}{5} \\ \hline \end{array}$ 14. $\begin{array}{r} 4\frac{3}{10} \\ + 2\frac{1}{2} \\ \hline \end{array}$

Subtract. Write each difference in simplest form.

15. $\frac{9}{10} - \frac{1}{10}$ 16. $\frac{7}{8} - \frac{1}{2}$ 17. $\frac{5}{8} - \frac{2}{4}$ 18. $\frac{10}{12} - \frac{1}{3}$ 19. $4\frac{6}{7} - \frac{6}{7}$

20. $\begin{array}{r} 6\frac{4}{5} \\ - 3\frac{2}{5} \\ \hline \end{array}$ 21. $\begin{array}{r} 8\frac{7}{8} \\ - 5\frac{1}{2} \\ \hline \end{array}$ 22. $\begin{array}{r} 3\frac{2}{5} \\ - 1\frac{1}{4} \\ \hline \end{array}$ 23. $\begin{array}{r} 4\frac{8}{9} \\ - 1\frac{2}{3} \\ \hline \end{array}$ 24. $\begin{array}{r} \frac{6}{9} \\ - \frac{1}{3} \\ \hline \end{array}$

Solve.

25. Audrey plans to make applesauce. The recipe calls for $1\frac{1}{4}$ pounds of apples. She has $\frac{3}{5}$ of a pound. How much more does Audrey need?

26. Andy's dog has a litter of 12 puppies that are yellow, brown, and black. If $\frac{1}{3}$ of the puppies are black and $\frac{1}{2}$ of the puppies are yellow, how many puppies are brown?

27. Adam weighed three kittens. The calico kitten was $2\frac{3}{4}$ pounds. The striped kitten was $3\frac{1}{3}$ pounds. And, the black kitten was $4\frac{1}{2}$ pounds. How much did the kittens weigh altogether?

Signing Math

American Sign Language is the third most-used language in the United States. It is the main sign language used in deaf communities in the United States. Somewhere between 500,000 to 2,000,000 people in the United States use this as their main way of communicating. People also use it in the English-speaking parts of Canada, parts of Mexico, several African countries, and in the countries of Hong Kong, Singapore, Philippines, Haiti, Puerto Rico, and Dominican Republic. There are hand signals for letters and words, and for numbers and math concepts too.

Try your hand at decoding some math-related American Sign Language hand signals. Use the following chart. Then study the following problems to sign some numbers or to unlock the mystery of the math signs.

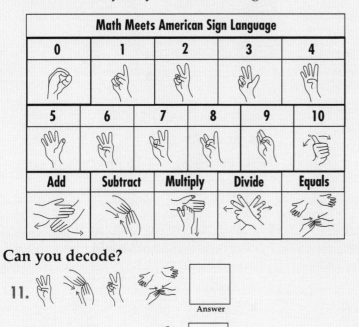

Can you decode?

11.
12.
13.
14.

Answer

Can you sign?

1. your birthday
2. your birth year
3. your zip code
4. your age
5. the current year
6. your phone number
7. the page number of this book
8. $10 \div 3$
9. 3×3
10. $(7 + 2) - 5$

CUMULATIVE TEST

1. Choose the best estimate for 45×82.
   a. 400
   b. 3,200
   c. 4,000
   d. none of the above

2.
$$\begin{array}{r} 27{,}892 \\ + 45{,}677 \\ \hline \end{array}$$
   a. 62,469
   b. 72,469
   c. 73,569
   d. none of the above

3. $1{,}242 - 576$
   a. 666
   b. 1,334
   c. 1,818
   d. none of the above

4. Which fraction is greater than $\frac{4}{6}$?
   a. $\frac{1}{4}$
   b. $\frac{1}{2}$
   c. $\frac{2}{3}$
   d. none of the above

5. $\frac{5}{8} + \frac{1}{8}$
   a. $\frac{6}{16}$
   b. $1\frac{17}{40}$
   c. $1\frac{3}{4}$
   d. none of the above

6. $7\overline{)4{,}226}$
   a. 63 R5
   b. 603 R5
   c. 6,003 R5
   d. none of the above

7. Mrs. Green bought 240 tomato plants. She planted 87 yesterday and 54 today. How many more must she plant?
   a. 99 plants
   b. 153 plants
   c. 327 plants
   d. none of the above

8. What is this figure?

   R ●————————● S

   a. line segment RS
   b. angle RS
   c. line RS
   d. none of the above

9. What is the area of the colored part of the rectangle?

   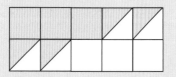

   a. 2 square units
   b. 5 square units
   c. 7 square units
   d. none of the above

10. Calvin bought two albums at $7.99 each. How much change should he receive from $20.00?
    a. $5.02
    b. $12.99
    c. $4.02
    d. $11.99

Adding and Subtracting Decimals

Gates Blair
Calvert Day School

12.1 Decimal Place Value: Tenths and Hundredths

Objective: to write tenths and hundredths as fractions and decimals

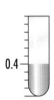

A simple invention used to measure the amount of rain is the rain gauge. This rain gauge has $\frac{4}{10}$ of an inch of water in it. This fraction can be written as a decimal. A decimal is a way of expressing fractions that have denominators of 10, 100, 1,000, and so on.

Fraction	Decimal
$\frac{4}{10}$ → four-tenths →	0.4

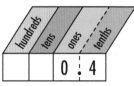

The place-value chart can be extended to show four-tenths.

More Examples

$\frac{4}{100}$ → four-hundredths → 0.04

$\frac{23}{100}$ → twenty-three hundredths → 0.23

$1\frac{23}{100}$ → one and twenty-three hundredths → 1.23

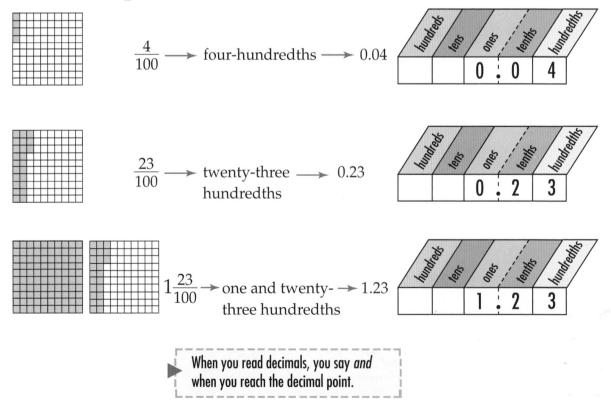

> When you read decimals, you say *and* when you reach the decimal point.

TRY THESE

Write a fraction or a mixed number and a decimal to name each colored part.

1.   2.   3.

Write each decimal in words and as a fraction or a mixed number.

4. hundreds | tens | ones | tenths | hundredths
    0 : 5 6

5. hundreds | tens | ones | tenths | hundredths
    2 9 : 3 0

6. hundreds | tens | ones | tenths | hundredths
    2 7 9 : 8 7

Exercises

Write the length of each red line segment as a decimal.

1.
```
<--|++++++++|++++++++|-->
   0        1        2
```

2.
```
<--|++++++++|++++++++|-->
   0        1        2
```

Write each fraction or mixed number as a decimal.

3. $\frac{5}{10}$ 4. $\frac{63}{100}$ 5. $22\frac{9}{10}$ 6. $3\frac{12}{100}$ 7. $31\frac{4}{10}$ 8. $51\frac{20}{100}$

Name the place-value position for the red digit in each number.

9. 0.3   10. 21.06   11. 23.5   12. 31.57   13. 49.3

PROBLEM SOLVING

Use the chart to solve.

14. Write in words and a fraction the least amount of rainfall in the world.

15. Write in words the least amount of rainfall in the United States. *Note:* There are two ways.

16. Write in words the greatest amount of rainfall in the world.

Yearly Rainfall Records	
Greatest in World (Hawaii)	1,168.47 cm
Least in World (Chile)	0.12 cm
Least in U.S. (California)	4.60 cm

12.2 Thousandths

Objective: to write thousandths as fractions and decimals

This block to the right shows the fraction $\frac{4}{1,000}$.

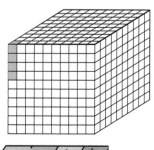

Since it has a denominator of 1,000, it can be written as a decimal.

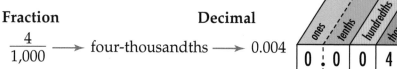

Fraction		**Decimal**
$\frac{4}{1,000}$ →	four-thousandths →	0.004

ones	tenths	hundredths	thousandths
0	0	0	4

The place-value chart can be extended to show four-thousandths.

Another Example

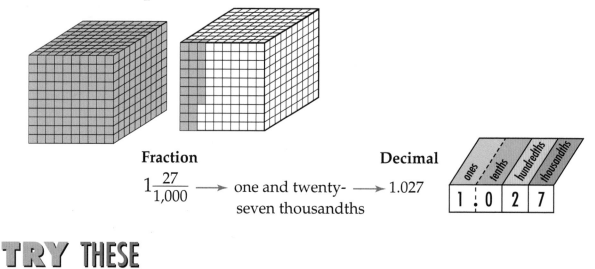

Fraction		**Decimal**
$1\frac{27}{1,000}$ →	one and twenty-seven thousandths →	1.027

ones	tenths	hundredths	thousandths
1	0	2	7

TRY THESE

Write each decimal in words and as a fraction or a mixed number.

1. 0.259　　　**2.** 19.908　　　**3.** 3.762　　　**4.** 10.003

Exercises

Write as a decimal.

1. 450 out of 1,000

2. 67 out of 1,000

3. 99 out of 1,000

4. $\dfrac{7}{1,000}$

5. $\dfrac{3}{1,000}$

6. $\dfrac{34}{100}$

7. $\dfrac{3}{10}$

8. $\dfrac{56}{1,000}$

9. $\dfrac{43}{100}$

10. $\dfrac{39}{100}$

11. $\dfrac{9}{10}$

12. $\dfrac{929}{1,000}$

Name the place-value position for the red digit in each number.

13. 3.457

14. 13.723

15. 107.51

16. 5.329

17. 14.771

18. 38.78

PROBLEM SOLVING

19. Ted ran a race in eleven and seven-tenths seconds. How would the timekeeper show Ted's time on a scorecard?

20. Kara found a treasure chest that weighed nineteen and two hundred forty-seven thousandths kilograms. Write its weight as a decimal.

21. A bumblebee is about two and sixty-three hundredths centimeters long. Write its length as a decimal.

MIXED REVIEW

Complete.

22. 4 cups = ■ pints

23. 2 cups = ■ pints

24. 16 cups = ■ pints

25. 2 pints = ■ quarts

26. 8 pints = ■ quarts

27. $8\frac{1}{2}$ cups = ■ quarts

12.3 More Fractions, Mixed Numbers, and Decimals

Objective: to practice with relating fractions and mixed numbers to decimals

Della set the trip odometer on her mother's car to see how far it was from her house to her school. Her mother told her it was $6\frac{3}{4}$ miles. What did the trip odometer read?

To find out what the trip odometer read, Della needed to find an equivalent fraction with a denominator of 10, 100, or 1,000. Then she can write the fraction as a decimal.

Since 4 does not go evenly into 10, she tried 100.

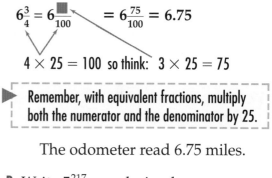

$6\frac{3}{4} = 6\frac{\blacksquare}{100} = 6\frac{75}{100} = 6.75$

$4 \times 25 = 100$ so think: $3 \times 25 = 75$

▶ Remember, with equivalent fractions, multiply both the numerator and the denominator by 25.

More Examples

The odometer read 6.75 miles.

A. Write $7\frac{1}{5}$ as a decimal.

$7\frac{1}{5} \times \frac{2}{2} = 7\frac{2}{10} = 7.2$

B. Write $5\frac{217}{250}$ as a decimal.

$5\frac{217}{250} \times \frac{4}{4} = 5\frac{868}{1,000} = 5.868$

TRY THESE

Copy and complete. Write each fraction as a decimal.

1. $\frac{1}{2} \times \frac{\blacksquare}{\blacksquare} = \frac{\blacksquare}{10} = \blacksquare$

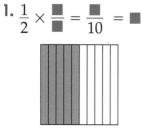

2. $\frac{10}{50} \times \frac{2}{2} = \frac{\blacksquare}{\blacksquare} = \blacksquare$

Exercises

Write each mixed number as a decimal.

1. $3\frac{1}{2}$

2. $4\frac{1}{4}$

3. $16\frac{2}{5}$

4. $11\frac{7}{20}$

5. $13\frac{14}{25}$

6. $25\frac{25}{200}$

7. $1\frac{3}{100}$

8. $19\frac{8}{250}$

9. $22\frac{319}{500}$

★ 10. $120\frac{8}{125}$

PROBLEM SOLVING

11. Ted ran a race in eleven and $\frac{1}{25}$ seconds. If the timekeeper wrote the time as a decimal on his scorecard, what would Ted's time look like?

12. Sean brought $5\frac{1}{5}$ containers of ice cream to the party. Write that amount as a decimal.

TEST PREP

13. Eliza was working on a 500-piece puzzle. She has found 81 pieces that fit together so far. What decimal tells the amount of the puzzle she has together?

   a. 0.81   **b.** 0.081   **c.** 0.162   **d.** 1.62

MIND BUILDER

Relations

A meterstick is marked into 100 centimeters. A centimeter equals 0.01 meters. Each centimeter is marked into 10 millimeters. A millimeter equals 0.001 meters.

100 cm = 1 m
1 cm = 0.01 m
1,000 mm = 1 m
1 mm = 0.001 m

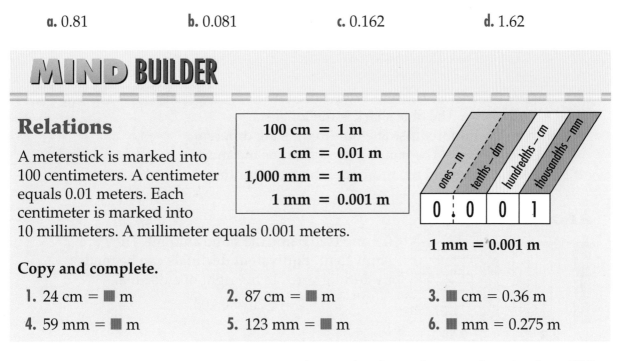

1 mm = 0.001 m

Copy and complete.

1. 24 cm = ■ m

2. 87 cm = ■ m

3. ■ cm = 0.36 m

4. 59 mm = ■ m

5. 123 mm = ■ m

6. ■ mm = 0.275 m

12.4 Comparing and Ordering Decimals

Objective: to compare and order decimal numbers

At the shark tank, there are several kinds of sharks. The bull shark is 11.245 feet long; the blue shark is 11.960 feet long; and the hammerhead shark is 11.231 feet long. Which shark is the longest? Which shark is the shortest?

Bull shark: 11.245 feet   Blue shark: 11.960 feet   Hammerhead shark: 11.231 feet

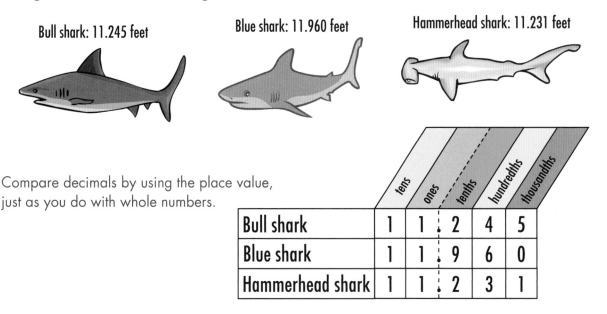

Compare decimals by using the place value, just as you do with whole numbers.

	tens	ones	tenths	hundredths	thousandths
Bull shark	1	1	2	4	5
Blue shark	1	1	9	6	0
Hammerhead shark	1	1	2	3	1

Start with the greatest place-value position.
- Compare the tens. The tens are the same.
- Compare the ones. The ones are the same.
- Compare the tenths. The tenths are different.
  0.9 > 0.2 ⟶ The blue shark is the longest.
- Compare the hundredths. The hundredths are different.
  0.03 < 0.04 ⟶ The hammerhead shark is the shortest.
- If needed, you would next compare the thousandths.

Another Example

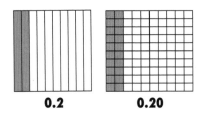

0.2   **0.20**

0.2 and 0.20 name the same amount. They are equivalent. **Equivalent decimals** can be made by adding zeros to the right of a decimal.

$$0.2 = 0.20 = 0.200$$

TRY THESE

Compare using <, >, or =.

1. 2.35 ● 2.53   **2.** 24.872 ● 24.641   **3.** 1.090 ● 1.09   **4.** 109.90 ● 109.98

Compare using <, >, or =.

1. 3.4 ● 3.45   **2.** 5.782 ● 5.874   **3.** 67.87 ● 67.27   **4.** 52.152 ● 52.153

Order from least to greatest.

5. 0.7 0.1 0.3 0.4   **6.** 12.48 12.53 12.87 11.99

7. 45.716 45.222 45.671   **8.** 4.123 4.132 4.321 4.312

PROBLEM SOLVING

Use the chart to solve. In problems 9–12, which is the longer whale?

9. humpback or bowhead

10. orca or minke

11. bowhead or gray

12. minke or humpback

13. Of all the whales, which is the longest?

14. Of all the whales, which is the shortest?

Length of Whales	
Humpback	50.4 feet
Gray	50.5 feet
Bowhead	50.61 feet
Orca	28.6 feet
Minke	28.34 feet

CONSTRUCTED RESPONSE

★ **15.** Write the decimal 123.456 as a fraction and in words. Then write a decimal that is greater but less than 124.00. Explain how you know your number is greater than the one given.

12.5 Comparing and Ordering Fractions and Decimals

Objective: to compare and order decimals and fractions involving tenths and hundredths

Tek and Lou were being backyard scientists one Saturday, experimenting with some bubble solution recipes they had. Tek's recipe called for 0.5 liters of dishwashing liquid. Lou's recipe called for $\frac{4}{10}$ of a liter of dishwashing liquid. Whose recipe needed more?

To find the answer, first you need to remember that fractions can be written as decimals, and decimals can be written as fractions.

$$0.5 \text{ liters} = \frac{5}{10} \text{ liters} \qquad \frac{4}{10} \text{ liters} = 0.4 \text{ liters}$$

Using this information, you can compare these numbers by looking at either the fractions or the decimals.

$$0.5 > 0.4 \text{ or } \frac{5}{10} > \frac{4}{10}$$

With both, 5 parts out of 10 is more than 4 parts out of 10. Tek's recipe needs more dishwashing liquid.

Another Example

Tek's recipe needs 0.060 liters of glycerin. Lou's recipe needs 0.06 liters of glycerin. Who needs more?

$$0.060 = \frac{60}{1,000} = \frac{6}{100}$$

$$0.06 = \frac{6}{100}$$

They both need the same amount.

TRY THESE

Compare using <, >, or =.

1. $\frac{5}{10}$ ● 0.47

2. Tek needs 4.4 liters of water for his bubbles. Lou needs $4\frac{1}{4}$ liters of water. Who needs less?

$$4.4 \; ● \; 4\frac{1}{4}$$

Exercises

Compare using <, >, or =.

1. 0.9 ● $\frac{88}{100}$

2. $\frac{75}{100}$ ● 0.75

3. 0.68 ● $\frac{6}{10}$

4. 0.12 ● 12

5. $\frac{37}{100}$ ● 3.7

6. $\frac{5}{100}$ ● 0.5

Order from least to greatest.

7. $\frac{8}{100}$ $\frac{7}{10}$ 0.055

8. 0.43 $\frac{41}{100}$ $\frac{4}{10}$

9. $\frac{1}{2}$ $\frac{55}{100}$ 0.034

PROBLEM SOLVING

10. Gino lives 0.68 miles from the park. Stephanie lives $\frac{7}{10}$ of a mile from the park. Who lives closer to the park?

11. At the ball game, 0.47 of the people were men, $\frac{23}{100}$ were women, and $\frac{3}{10}$ were children. Order these groups from least to greatest.

TEST PREP

12. Which set of numbers is in order from greatest to least?

    a. $\frac{4}{10}$ 0.060 $\frac{77}{100}$
    b. 0.8 $\frac{5}{100}$ 0.35
    c. 0.98 $\frac{3}{10}$ $\frac{48}{100}$
    d. $\frac{85}{100}$ 0.66 $\frac{3}{10}$

13. Hal and Belle each had marble collections of the same size. Hal's collection was $\frac{45}{100}$ blue marbles. Belle's collection was 0.4 blue ones. Who had more blue marbles?

    a. Hal   b. Belle   c. They had the same amount.

12.6 Problem-Solving Strategy: Using Venn Diagrams

Objective: to solve problems using Venn diagrams

The chart at the right shows information about club membership at Parkview School. Sometimes chart information is easier to see if it is put into a special diagram called a **Venn diagram**.

Club Membership at Parkview School	
Number of Members	Clubs
8	Computer and Invention
19	Invention Only
18	Computer Only

1. READ
You know how many students are in each club.
You want to organize this information on a Venn diagram.

2. PLAN

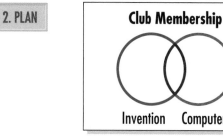

- Each circle represents a club.

- Students who are members of both clubs are shown where the circles overlap.

3. SOLVE
Start with the number of students in *both* clubs. Place an 8 in the area that is in *both* circles.

Then place the numbers of students who are only in *one* club.

The completed diagram looks like this.

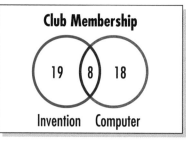

4. CHECK
Compare the chart and the diagram.
Be sure both show the same information.

TRY THESE

Use the Venn diagram on the previous page to solve.

1. How many students are only in the computer club?

2. Which circle represents the computer club?

3. How many members does the computer club have?

4. What is the total number of students in the invention club?

5. Which circle represents the invention club?

6. How many students are in a school club?

Solve

Use the Venn diagram to solve.

1. What does the red circle represent?

2. How many students went swimming?

3. How many students did not go swimming?

4. How many students were at the picnic?

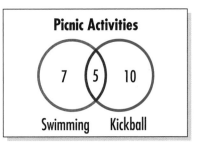

Picnic Activities

7 | 5 | 10

Swimming   Kickball

CONSTRUCTED RESPONSE

5. There were 30 people at the Sunnybrooke Horse Farm. Of those people, 12 people signed up for horseback riding lessons, and 9 people signed up for the Proper Care and Treatment of Horses class. The rest of the people signed up for both the lesson and the class. Draw a Venn diagram and explain why you placed the numbers where you did.

6. At Deno's Diner on Saturday, 75 chicken meals were ordered for lunch. Deno drew up this Venn diagram to show the owner what meals were ordered. Explain Deno's Venn diagram.

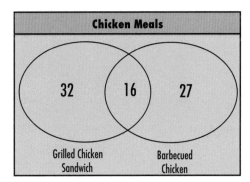

Chicken Meals

32 | 16 | 27

Grilled Chicken   Barbecued
Sandwich   Chicken

What Are the Letters?

Lenny watched his baby sister, Jenny, stack her alphabet blocks. The three remaining blocks are identical, but they are positioned in different directions. What letters are on the faces opposite letters **B**, **S**, and **F**? (Give your answer in the same order.)

CUMULATIVE REVIEW

Compute.

1. $\begin{array}{r} 300 \\ \times 7 \\ \hline \end{array}$

2. $\begin{array}{r} \$157.84 \\ -98.75 \\ \hline \end{array}$

3. $\begin{array}{r} 83 \\ \times72 \\ \hline \end{array}$

4. $\begin{array}{r} 7{,}638 \\ +9{,}557 \\ \hline \end{array}$

5. $\begin{array}{r} 259 \\ \times48 \\ \hline \end{array}$

6. $3{,}784 \div 5$

7. $7{,}403 - 3{,}765$

8. $37{,}849 + 48{,}674$

9. $14 \times \$3.75$

10. $879 \div 6$

11. $8{,}322 - 3{,}956$

Compare using <, >, or =.

12. $0.8 \bullet 1.08$

13. $0.5 \bullet 0.50$

14. $3.2 \bullet 3.1$

Use the figures to the right to answer the following questions.

15. Which figure is a cone?

16. Which figure is a pentagon?

17. Which figure is a rectangular prism?

Write a fraction for the colored part.

18.

19.

20.

Find the area.

21.

4 ft

4 ft

3 ft

2 ft

2 ft 1 ft

Find the perimeter.

22.

19 m

11 m

12 m

15 m

23 m

Find the volume.

23.

1 cm

3 cm

4 cm

Write as a decimal.

24. 11 out of 100

25. 2 out of 10

26. 37 out of 100

27. $\dfrac{68}{100}$

28. $2\dfrac{17}{100}$

29. $\dfrac{6}{10}$

30. $\dfrac{42}{100}$

12.7 Rounding Decimals

Objective: to round and estimate with decimals

Kora and her sister discovered a small park 0.9 miles from their new home. To the nearest whole number, how far is the park from their home?

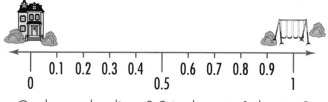

On the number line, 0.9 is closer to 1 than to 0.

To the nearest whole number, 0.9 rounds to 1. The park is about 1 mile from their home.

Round decimals like you round whole numbers.

| ▶ Round up if the digit to the right is 5, 6, 7, 8, or 9. | ▶ The digit remains the same if the digit to the right is 0, 1, 2, 3, or 4. |

More Examples

A. Round 0.2 to the nearest whole number.

Answer: 0   ▶ Since the digit to the right of 0 is 2, the nearest whole number is 0.

B. Round 1.29 to the nearest tenth.

Answer: 1.3   ▶ Since the digit to the right of 2 is 9, round 2 tenths up to 3 tenths.

Choose the letter of the best estimate. (*Hint:* Round each number to the nearest whole number *before* adding or subtracting mentally.)

C. 0.2 + 0.7 =   a. 0   b. 1   c. 2

0 + 1   ▶ Since the sum is close to 1, the best estimate is **b**.

D. 0.91 − 0.71   a. 0   b. 1   c. 2

1 − 1   ▶ Since the difference is close to 0, the best estimate is **a**.

TRY THESE

Use the number line to round to the nearest tenth.

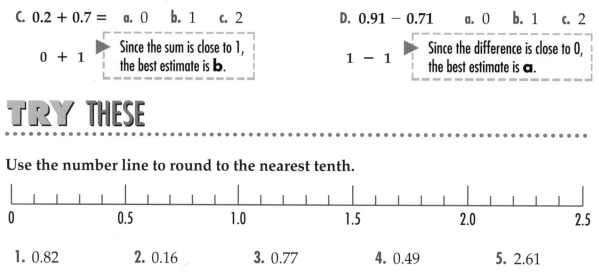

1. 0.82   **2.** 0.16   **3.** 0.77   **4.** 0.49   **5.** 2.61

Exercises

Round to the nearest tenth.

1. 0.22   **2.** 0.84   **3.** 0.69   **4.** 0.09   **5.** 1.14   **6.** 1.93

7. 0.71   **8.** 0.38   **9.** 0.59   **10.** 0.22   **11.** 1.31   **12.** 2.35

Choose the letter of the best estimate.

13. $0.9 + 0.1$ **a.** 0   **b.** 1   **c.** 2   **14.** $0.6 - 0.3$ **a.** 0   **b.** 1   **c.** 2

15. $0.77 + 0.88$ **a.** 0   **b.** 1   **c.** 2   **16.** $0.87 - 0.54$ **a.** 0   **b.** 1   **c.** 2

PROBLEM SOLVING

17. Kent Jones ran 1.4 miles one week and 3.8 miles the next week. Estimate the total number of miles Kent ran.

18. Cindy has 24 packs of football cards. There are 15 cards in each pack. How many football cards does Cindy have?

MID-CHAPTER REVIEW

Write a decimal for the colored area.

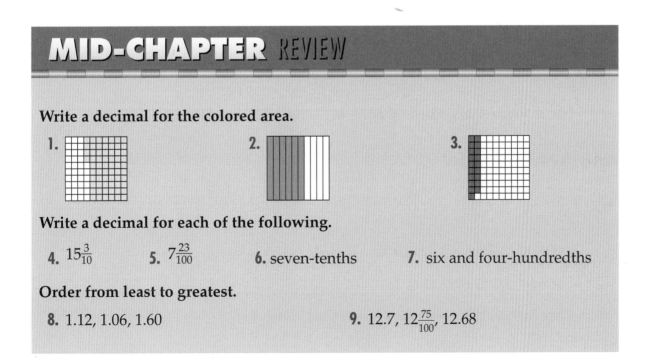

1.   **2.**   **3.**

Write a decimal for each of the following.

4. $15\frac{3}{10}$ **5.** $7\frac{23}{100}$ **6.** seven-tenths   **7.** six and four-hundredths

Order from least to greatest.

8. 1.12, 1.06, 1.60   **9.** 12.7, $12\frac{75}{100}$, 12.68

12.8 Estimating Decimal Sums and Differences

Objective: to estimate decimal sums and differences using rounding

Fred and his family were planning their family vacation. To help them decide where to go, they studied a distance mileage chart. One of their ideas was to drive from Baltimore to Chicago and then back again, but they did not want to travel more than 1,500 miles total.

As they were figuring out how far a round-trip between Baltimore and Chicago was, they did not need an exact number of miles. An estimate would give them enough information.

Driving Distances (in miles)

	Baltimore	Chicago	Cincinnati	St. Louis	Tampa
Baltimore		703.34	519.37	879.96	957.03
Chicago	703.34		297.28	297.27	1,174.37
Cincinnati	519.37	297.28		351.32	924.84
St. Louis	879.96	297.27	351.32		1,010.83
Tampa	957.03	1,174.37	924.84	1,010.83	

First, they needed to round how far it was from Baltimore to Chicago to the nearest mile.

> 703.34 miles rounds to 703 miles.

Since they were planning to drive to Chicago and then back again, they needed to add 703 twice.

> **703 + 703 = 1,406 miles**   This trip would be less than 1,500 miles.

Another Example

From Tampa to St. Louis, it is 1,010.83 miles. About how much less than 1,200 miles is this trip?

> **THINK:** 1,010.83 rounds to 1,011.
>
> Subtract. **1,200 − 1,011 = 189**
>
> **This trip is about 189 miles less than 1,200 miles.**

TRY THESE

Estimate using rounding.

1. $17.38 + $275.87

2. 59.1 m − 29.6 m

3. 241.45 lb + 75.9 lb

Exercises

Estimate using rounding.

1. $8.94
       + $73.27

2.   129.4
       − 39.29

3.   392.39
       + 287.63

4.   1,290.22
       − 192.75

5. $280.4
       + $270.74

PROBLEM SOLVING

Use the Driving Distance Mileage Chart on the previous page to solve.

6. A trip from Baltimore to Chicago is about how many miles less than 1,000 miles?

7. About how much farther is it from Baltimore to Tampa than it is from Baltimore to Chicago?

8. Marci drove from St. Louis to see her grandmother in Baltimore. On the way back to St. Louis she stopped in Cincinnati to visit her best friend. Estimate the total number of miles she drove by the time she got back home to St. Louis.

9. Beau traveled from Chicago to Cincinnati to St. Louis and then back to Chicago. About how far did he drive?

CONSTRUCTED RESPONSE

★ **10.** Study the mileage chart on the previous page. Estimate to find the longest trip you can make traveling through four cities. Explain your reasoning.

12.9 Adding and Subtracting Decimals

Objective: to add and subtract decimals

In 1893, Thomas Edison invented the first continuous roll of film. This and other discoveries led to the film used for movies today. How long will it take to show both of these films?

To find the total, add 26.5 and 15.

▶ Line up the place-value positions. Write 15 as 15.0 to make adding easier.

$$\begin{array}{r} 26.5 \\ + \ 15 \end{array}$$

	tens	ones	tenths
26.5	2	6	5
+ 15	+1	5	0

Zero has been written in as a placeholder. The value of the number does not change.

THINK
Estimate
26.5 + 15 as 50.

30 + 20 = 50

	Step 1	Step 2	Step 3
	Add the tenths.	Add the ones.	Add the tens.
	$\begin{array}{r} 26.5 \\ + 15.0 \\ \hline .5 \end{array}$	$\begin{array}{r} 1 \\ 26.5 \\ + 15.0 \\ \hline 1.5 \end{array}$	$\begin{array}{r} 1 \\ 26.5 \\ + 15.0 \\ \hline 41.5 \end{array}$
	Put the decimal point in the answer.	Rename 11 ones as 1 ten and 1 one.	

It will take 41.5 minutes to show both films.

More Examples

▶ Is the answer reasonable?

A.
$$\begin{array}{r} {}^{6\ \ 16\,12} \\ \cancel{7.72} \\ - \ 3.84 \\ \hline 3.88 \end{array}$$

Check:
$$\begin{array}{r} {}^{1\ \ 1} \\ 3.84 \\ + \ 3.88 \\ \hline 7.72 \end{array}$$

B.
$$\begin{array}{r} {}^{2\,15\ \ 11\,15} \\ \cancel{36.25} \\ - \ 17.36 \\ \hline 18.89 \end{array}$$

TRY THESE

Add or subtract.

1. $\begin{array}{r} 3.6 \\ + 2.3 \end{array}$

2. $\begin{array}{r} 7.5 \\ - 4.3 \end{array}$

3. $\begin{array}{r} 53.4 \\ - 31.8 \end{array}$

4. $\begin{array}{r} 45.4 \\ + 57.9 \end{array}$

Exercises

Add or subtract.

1. 0.56
    + 0.75

2. 24.6
    − 13.7

3. 5.81
    − 2.63

4. 43.42
    + 29.35

5. $34.26
    − 17.08

6. $38.79
    − 15.65

7. 2.56
    + 9.37

8. 97.35
    + 56.91

9. $64.98
    + 49.99

10. 52.07
    − 45.93

11. 52.07 − 45.93

12. Subtract $2.36 from $8.00.

PROBLEM SOLVING

Use the bar graph to solve.

13. How much greater was Burt's top speed on Saturday than on Friday?

14. How much faster did Burt drive than Ralph on Friday?

★15. What was Ralph's average top speed for the 3 days?

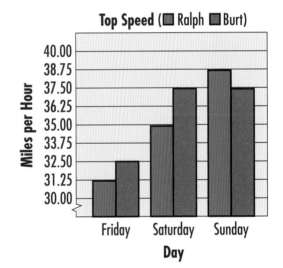

MIXED REVIEW

Find the area.

16.
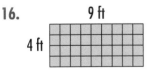
9 ft
4 ft

17.
6 m
6 m

18.
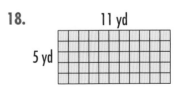
11 yd
5 yd

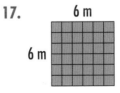

12.10 Problem-Solving Application: Using Measurements

Objective: to solve measurement problems that involve decimals

Maple Elementary School has two winners in the county math contest. A camera crew from a local TV studio travels 8.7 kilometers to tape the school assembly. After driving 2.1 kilometers, the camera crew stops at a gasoline station. How far is it from the gasoline station to the school?

1. READ	You need to know the distance from the gasoline station to the school. You know the distances from the TV studio to the school and from the TV studio to the gasoline station.

2. PLAN	The camera crew has traveled 2.1 of the total 8.7 kilometers. So, subtract to find the distance from the gasoline station to the school. First estimate. Write an equation.

$$n = 8.7 - 2.1$$

3. SOLVE	Estimate $8.7 - 2.1$ to be $9 - 2$, or 7.

$$\begin{array}{r} 8.7 \\ -\ 2.1 \\ \hline 6.6 \end{array}$$ total distance
distance to gasoline station

It is 6.6 kilometers from the gasoline station to the school.

4. CHECK	Use addition to check subtraction. The estimate of about 7 miles is reasonable.

$$\begin{array}{r} 6.6 \\ +\ 2.1 \\ \hline 8.7 \end{array}$$ The answer checks.

TRY THESE

Solve.

1. Carol paid $11.55 for gasoline on Wednesday and $9.40 on Friday. How much did she spend for gasoline that week?

2. Mrs. Walters used 1.9 liters of paint to paint a picnic table. How much paint is left in the can?

Paint 3.5 L

Solve...

1. Bob had 12.3 meters of rope. He used 2.5 meters. How much rope does he have left?

2. Last fall, Ted weighed 37.4 kg. He gained 3.7 kg. How much does Ted weigh now?

The camera crew from the TV studio kept a record of the kilometers traveled. Use the chart to solve problems 3–8.

Distance Traveled by the Camera Crew						
Day	Monday	Tuesday	Wednesday	Thursday	Friday	Saturday
Kilometers	84.8	58.71	66.93	72.45	84.6	92.5

3. Find the total number of kilometers traveled on Monday and Tuesday.

4. How much farther did the crew travel on Saturday than on Friday?

5. Find the total distance traveled on Wednesday, Thursday, and Friday.

6. On which day did the crew travel the greatest number of kilometers?

7. On which day did the crew travel the least number of kilometers?

8. How much farther did the crew travel on Monday than on Tuesday?

MIND BUILDER

Logical Thinking

Copy and complete.

1. $\begin{array}{r} 5.7\blacksquare \\ + 1.\blacksquare6 \\ \hline \blacksquare.59 \end{array}$

2. $\begin{array}{r} 17.\blacksquare4 \\ - 9.5\blacksquare \\ \hline \blacksquare.83 \end{array}$

3. $\begin{array}{r} \blacksquare5.0\blacksquare \\ + 6.64 \\ \hline 3\blacksquare.64 \end{array}$

4. $\begin{array}{r} 6\blacksquare.2 \\ - \blacksquare1.\blacksquare \\ \hline 21.4 \end{array}$

5. $\begin{array}{r} 3\blacksquare.78 \\ + 11.\blacksquare1 \\ \hline \blacksquare1.8\blacksquare \end{array}$

CHAPTER 12 REVIEW

LANGUAGE and CONCEPTS

Choose the correct word to complete each sentence.

1. Read the decimal point as (*and*, *or*, *of*, *for*).

2. The answer in the subtraction of decimals is called the (sum, difference, product, denominator).

3. Ones divided into ten equal parts are called (tenths, hundredths, tens, numerators).

4. Tenths divided into ten equal parts are called (tenths, hundredths, tens, numerators).

5. Always line up the digits in the same (number, equal, place-value, numerator) positions before adding or subtracting decimals.

★ 6. You can rename sixteen-hundredths as one-(one, ten, tenth, hundredth) and six-hundredths.

SKILLS and PROBLEM SOLVING

Write a decimal for the colored area. (Sections 12.1–12.3)

7.

8.

9.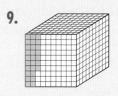

Write a decimal for each of the following. (Sections 12.1–12.3)

10. $8\frac{5}{10}$

11. two and six-hundredths

12. $17\frac{871}{1,000}$

13. nine-tenths

Compare using <, >, or =. (Sections 12.4–12.5)

14. 0.8 ● 0.6

15. $1\frac{72}{100}$ ● 1.72

16. 5.61 ● 5.6

17. 0.9 ● 0.82

18. $\frac{4}{10}$ ● 0.4

19. $3\frac{27}{100}$ ● 3.3

Round to the nearest tenth. (Section 12.7)

20. 0.67
21. 0.19
22. 0.91
23. 1.23
24. 2.90

Estimate. (Section 12.8)

25.   0.25
     + 5.63

26.   6.9
     − 5.75

27.   8.6
     − 1.2

28.   4.8
     + 7.9

29. $7.52
     − 0.83

Add or subtract. (Section 12.9)

30.   7.76
     − 4.17

31.   8.4
     − 2.25

32. $1.65
     + 0.79

33.   8.03
     − 5.5

34.   0.84
     + 6.16

35.   4.6
     + 5.01

36. $8.75
     − 0.50

37.   6.66
     − 3.37

38.   4.2
     + 8.80

39. $9.06
     − 5.69

Solve. (Section 12.10)

40. A farmer has 75.3 pounds of apples in two baskets. One basket weighs 32.8 pounds. What is the weight of the other basket?

41. One board is 2.73 meters long. Another board is 1.4 meters long. What is the total length of the two boards?

Use the Venn diagram to solve. (Section 12.6)

42. How many people went camping?

43. How many people went only to the zoo?

44. What is the total number of people who went to the zoo?

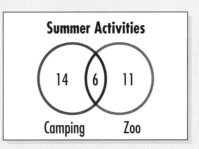

Summer Activities

14 (6) 11

Camping   Zoo

CHAPTER 12 TEST

Write as a decimal.

1. 57 out of 100

2. $2\frac{6}{10}$

3. $63\frac{247}{100}$

4. $32\frac{1}{10}$

5. fifty-one hundredths

6. seventy-one and four-thousandths

Compare using <, >, or =.

7. 0.6 ● 0.7

8. $\frac{89}{100}$ ● 0.98

9. 32.07 ● $23\frac{7}{10}$

Round to the nearest tenth.

10. 0.33

11. 0.69

12. 1.17

13. 2.54

14. 24.66

Estimate first. Then add or subtract.

15. $3.47 + 1.64

16. 24.03 − 22.98

17. 7.20 + 1.97

18. $62.08 − 27.90

19. 13.6 + 52.7

Solve.

20.

Rainfall (cm)	
July	15.71
August	10.35

Estimate the difference in the rainfall between the two months.

21. Mikki has walked 0.50 kilometers on her way to the movies. If the theater is 1.25 kilometers from her house, how much farther must she walk?

22. Mrs. Raymond bought 2.78 pounds of peaches and 3.9 pounds of grapes. How many total pounds of fruit did she buy?

23. Sara jogged 1.50 kilometers. Bruce jogged 0.75 kilometers more than Sara. How far did Bruce jog?

Use the Venn diagram to solve.

24. How many students are only in the drama club?

25. What is the total number of students in the chorus?

26. How many students are in both activities?

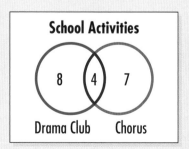

School Activities

8 4 7

Drama Club   Chorus

Net Detective

Remember, a net is a two-dimensional, flat pattern for a three-dimensional shape. It is what the shape would look like if you took it apart, unfolded it, and flattened it out.

For example, here is an octahedron, and here is its net if you took it apart and unfolded it:

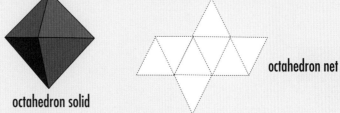

octahedron net

octahedron solid

Become a net detective using the following patterns. Can you find 10 nets that correctly fold to form a 6-sided cube?

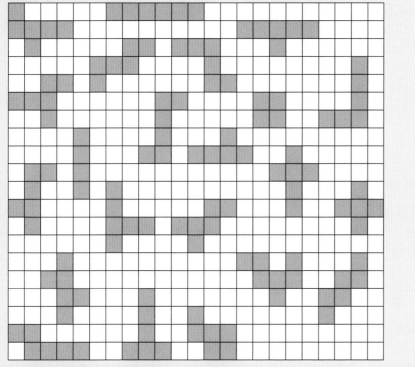

There are 11 ways to make a cube net. Can you create one more way?

CUMULATIVE TEST

1. 4)184

  a. 41
  b. 46
  c. 48
  d. none of the above

2. 8 m = ■ cm

  a. 80
  b. 800
  c. 8,000
  d. none of the above

3. Identify this angle.

  a. obtuse
  b. right
  c. acute
  d. none of the above

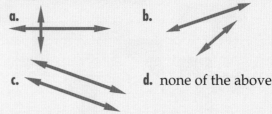

4. What fraction of the square is colored?

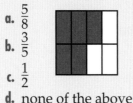

  a. $\frac{5}{8}$
  b. $\frac{3}{5}$
  c. $\frac{1}{2}$
  d. none of the above

5. $\frac{6}{7} + \frac{5}{7}$

  a. $\frac{11}{14}$
  b. $1\frac{4}{7}$
  c. $1\frac{11}{14}$
  d. none of the above

6. 32.40 − 17.28

  a. 15.12
  b. 25.12
  c. 25.28
  d. none of the above

7. There are 15 airplanes. Each airplane has 124 seats. How many seats are there in all?

  a. 1,860 seats
  b. 1,870 seats
  c. 1,880 seats
  d. none of the above

8. Which of the following shows parallel lines?

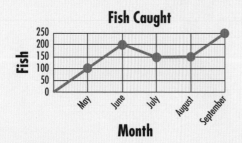

  d. none of the above

9. Find the average number of fish caught from May to September at Lake Jobe.

Fish Caught

  a. 150 fish
  b. 170 fish
  c. 200 fish
  d. none of the above

10. *If Joe weighs 50 pounds and Jill weighs 40 pounds, how much do they weigh in all?*

Which problem can be solved using the same operation as the problem shown above?

  a. Bill bought 6 apples at 35¢ each. How much did Bill spend?
  b. How much more does a 9-lb baby weigh than a 6-lb baby?
  c. Sue has 6 pears and 4 apples. How many pieces of fruit does she have?
  d. If there are 25 chairs in 4 rows, how many chairs are in each row?

A P P E N D I X

Mathematical Symbols

=	is equal to	°	degrees
%	percent	>	is greater than
<	is less than	⌐	right angle
≈	is approximately equal to	$\overrightarrow{AB}$	ray AB
$\overline{AB}$	line segment AB	$\overleftrightarrow{AB}$	line AB
△	triangle	∠	angle
⊥	is perpendicular to	‖	is parallel to

Metric System of Measurement

Prefixes kilo (k) = thousand
deci (d) = tenth
centi (c) = hundredth
milli (m) = thousandth

Length 1 centimeter (cm) = 10 millimeters (mm)
1 decimeter (dm) = 10 centimeters
1 meter (m) = 10 decimeters or 100 centimeters
1 kilometer (km) = 1,000 meters

Weight 1 kilogram (kg) = 1,000 grams (g)

Capacity 1 liter (L) = 1,000 milliliters (mL)

Customary System of Measurement

Length   1 foot (ft) = 12 inches (in.)
             1 yard (yd) = 3 feet or 36 inches
             1 mile (mi) = 1,760 yards or 5,280 feet

Weight   1 pound (lb) = 16 ounces (oz)
             1 ton = 2,000 pounds

Capacity   1 cup (c) = 8 ounces
             1 pint = 2 cups (c)
             1 quart (qt) = 2 pints
             1 gallon (gal) = 4 quarts

GLOSSARY

A

acute angle   **222**   An angle smaller than a right angle.

addend   **28**   A number that is added to another number. In the addition sentence 6 + 7 = 13, the addends are 6 and 7.

addition   **28**   The operation used to combine groups or to find a total.

$$5 + 6 = 11$$

analog clock   **210**   A clock that shows the time using hands on a dial.

angle   **262**   Two rays with a common endpoint.

area   **240**   The number of square units needed to cover a region.

Associative Property of Addition   **28**   The way in which addends are grouped does not change the sum.

$$(3 + 4) + 2 = 3 + (4 + 2)$$

Associative Property of Multiplication   **56** The way in which factors are grouped does not change the product.

$$(3 \times 4) \times 5 = 3 \times (4 \times 5)$$

average   **108**   See *mean*.

B

bar graph   **114**   A graph that is used to compare data in a visible way.

C

Celsius (°C)   **208**   The temperature scale used in the metric system. Water freezes at 0°C and boils at 100°C.

centimeter (cm)   **198**   A unit of length in the metric system.

$$100 \text{ centimeters} = 1 \text{ meter}$$

1 centimeter

circle graph   **131**   A graph that compares parts to a whole.

Commutative Property of Addition   **28**   The order of the addends does not change the sum.

$$7 + 2 = 9$$
$$2 + 7 = 9$$

Commutative Property of Multiplication   **56** The order of factors does not change the product.

$$7 \times 3 = 21$$
$$3 \times 7 = 21$$

compatible numbers   **168**   Numbers near the value of the numbers in the original problem that make the arithmetic easier. In division, compatible numbers are usually multiples of the divisors.

cone   **242**   A solid figure with one circular base and one curved surface.

congruent figures   **228** Figures that have the same size and shape.

constructed response   **16**   A descriptive answer to a problem that not only gives the solution, but also shows your thinking.

coordinates   **122**   The numbers in an ordered pair.

(2, 3)

cube 242 A rectangular prism that has six faces, each of which is a square.

cup (c) 192 A unit of capacity in the customary system of measurement.

1 cup = 8 ounces

customary system 190 A system of units of measure. The units of length are inches, feet, yards, and miles. The units of capacity are cups, pints, quarts, and gallons. The units of weight are ounces and pounds. Degree Fahrenheit is the unit used to measure temperature.

cylinder 243 A solid figure with two parallel congruent circular bases.

D

data 106 Data is another word for information.

decimal xl, 308 A number such as 0.3. Another way to write a fraction that has a denominator of 10, 100, 1,000, and so on.

decimeter (dm) 198 A unit of length in the metric system.

1 decimeter = 10 centimeters
10 decimeters = 1 meter

degree 208 A unit for measuring angles or temperature.

denominator 254 In the fraction $\frac{1}{4}$, the denominator is 4. It tells the number of objects or equal-sized parts.

difference 42 The answer to a subtraction problem.

$$11 - 6 = 5 \longleftarrow \text{difference}$$

digit 2 One of the following numerals: 0, 1, 2, 3, 4, 5, 6, 7, 8, 9. Digits are symbols used to name numbers.

digital clock 210 A clock that shows the time in number form.

Distributive Property 58 The product of a number and sum is equal to the sum of the products.

$$4 \times (3 + 2) = (4 \times 3) + (4 \times 2)$$

dividend 168 A number that is divided by another number. In the division sentence 24 ÷ 4 = 6, the dividend is 24.

divisible 163 A number that can be divided by another number equally with no remainder.

division 136 The operation used to separate a group into smaller groups of the same number or to find how many equal groups can be formed.

$$12 \div 3 = 4 \quad 3\overline{)12}^{\,4}$$

divisor 168 The number by which a dividend is divided. In the division sentence 24 ÷ 4 = 6, the divisor is 4.

double-bar graph 114 A graph that uses side-by-side bars to compare data.

E

edge 242 The line segment in which two faces of a solid figure meet.

elapsed time 210 The time between two given times.

equation 30 A mathematical sentence which contains an equal sign.

$$2 + 3 = 5$$

equivalent decimals 314 Decimals that name the same amount.

equivalent fractions 258 Two or more fractions that name the same number.

$$\frac{1}{3} \quad \frac{2}{6} \quad \frac{3}{9}$$

estimate 34 To find an approximate answer. The sum of 484 and 337 is about 800.

expanded form of a number **2** A form of a number that shows the value of each digit. The expanded form for 36 is 30 + 6.

F

face **242** Each flat surface of a solid figure.

fact family **xxviii, 134** The related number sentences for addition and subtraction (or multiplication and division) that use the same three numbers.

2, 3, 5	2 + 3 = 5	5 − 3 = 2
	3 + 2 = 5	5 − 2 = 3
4, 8, 32	4 × 8 = 32	32 ÷ 8 = 4
	8 × 4 = 32	32 ÷ 4 = 8

factor **60** Any number that you multiply. In the multiplication sentence 5 × 3 = 15, the factors are 5 and 3.

Fahrenheit (°F) **208** The temperature scale used in the customary system. Water freezes at 32°F and boils at 212°F.

flip **234** See *reflection*

foot (ft) **190** A unit of length in the customary system of measurement.

$$1 \text{ foot} = 12 \text{ inches}$$

fraction **254** A number for part of a whole or group, such as $\frac{1}{4}$.

function **124** A rule for finding output.

function table **124** A table that lists the input and output values of a particular function.

G

gallon (gal) **192** A unit of capacity in the customary system of measurement.

$$1 \text{ gallon} = 4 \text{ quarts}$$

geometry **259** The study of shapes.

gram (g) **202** A unit of weight in the metric system.

$$1000 \text{ grams} = 1 \text{ kilogram}$$

Grouping Property of Addition **28** See *Associative Property of Addition*.

Grouping Property of Multiplication **56** See *Associative Property of Multiplication*.

H

hexagon **224** A polygon with six sides.

I

Identity Property of Addition **28** When 0 is added to any number, the sum is that number.

$$0 + 5 = 5$$

Identity Property of Multiplication **56** When one factor is 1, the product is the same as the other factor.

improper fractions **266** Fractions that name numbers greater than or equal to 1, such as $\frac{7}{4}$ or $\frac{8}{8}$.

inch (in.) **190** A unit of length in the customary system of measurement.

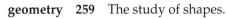

1 inch

independent event **276** A situation where one event does not affect the outcome of another, repeated event.

Ex. Each time you roll a die, one roll does not change the next roll.

input **124** The number that goes into a function; the number to which the function rule is applied.

intersecting lines **220** Lines in the same plane that cross.

K

key **118** The part of a display that defines abbreviations and symbols.

Key: 1|4 = 14

kilogram (kg) **202** A unit of mass in the metric system.

$$1 \text{ kilogram} = 1{,}000 \text{ grams}$$

kilometer (km) **198** A unit of length in the metric system.

$$1 \text{ kilometer} = 1{,}000 \text{ meters}$$

L

lattice multiplication **103** Using an array of boxes to find a product.

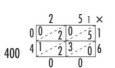

$$25 \times 16 = 400$$

least common denominator **294** The least common multiple of two or more denominators. The least common denominator of $\frac{2}{3}$ and $\frac{1}{4}$ is 12.

line **218** A never-ending straight path that extends in both directions.

line graph **116** A graph used to show change over a period of time.

line of symmetry **232** A line that separates a figure into two halves that match exactly.

line of symmetry

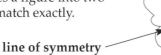

line plot **106** A display in which a mark is made over a number line to show each time a piece of data occurs.

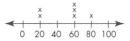

line segment **218** Two points and the straight path between them.

line symmetry **232** A figure has line symmetry if it can be folded so that both sides match exactly.

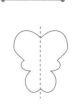

liter (L) **200** A unit of capacity in the metric system.

$$1 \text{ liter} = 1{,}000 \text{ milliliters}$$

M

magic square **51** A square array of numbers in which the sums of the rows, columns, and diagonals are equal.

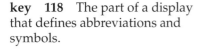

2	7	6
9	5	1
4	3	8

mass **202** The amount of matter of an object; a metric unit of measure.

mathematical expression **30** A translation of a word phrase or fragment into numbers and/or variables.

mean **108** Also called the average; the sum of a set of numbers divided by the number of addends.

median **108** The middle number when the data is listed in order.

Ex. 5, 7, 10, 11, 15 **10 is the median.**

meter (m) **198** A unit of length in the metric system.

$$1 \text{ meter} = 100 \text{ centimeters}$$

metric system **198** A system of units of measure that uses prefixes such as milli, centi, and kilo. The basic unit of length is meter. The basic unit of capacity is liter and the basic unit of mass is gram. Degree Celsius is the unit of measure for temperature.

mile (mi) **190** A unit of length in the customary system of measurement.

$$1 \text{ mile} = 5{,}280 \text{ feet}$$

milliliter (mL)   200   A unit of capacity in the metric system.

$$1,000 \text{ milliliters} = 1 \text{ liter}$$

millimeter   198   A unit of length in the metric system.

$$10 \text{ millimeters} = 1 \text{ centimeter}$$

mixed number   266   A number that has a whole number part and fraction part, such as $1\frac{3}{4}$.

mode   108   The number or item that appears most often in a set of data. Some sets of data have no mode or more than one mode.

Ex. 7, 3, 8, 2, 2, 9, 1   2 is the mode.

multiple   60   The product that occurs when a given number is multiplied by one of the whole numbers {0, 1, 2, 3, …}.

N

negative number   208   A number that is less than zero.

net   242   A two-dimensional pattern of a three-dimensional solid.

numerator   254   In the fraction $\frac{1}{4}$, the numerator is 1. It tells the number of objects or parts being considered.

O

obtuse angle   222   An angle larger than a right angle.

octagon   224   A polygon with eight sides.

operation symbols   30   Characters used to show the act of addition, subtraction, multiplication or division.

Ex. (+, −, ×, ÷)

Order Property of Addition   28   See *Commutative Property of Addition*.

Order Property of Multiplication   56   See *Commutative Property of Multiplication*.

ordered pair   122   A pair of numbers used to locate a point on a grid. The first number tells the horizontal position. The second number tells the vertical position.

(8, 9)

origin   126   The place where the *x*-axis meets the *y*-axis.

ounce (oz)   194   A unit of weight in the customary system of measurement.

$$16 \text{ ounces} = 1 \text{ pound}$$

outcome   274   A possible result.

output   124   The number that comes out of a function; the result of applying the function rule to the input.

P

palindrome   33   A number, word, or sentence that uses the same digits forward and backward.

363   23732   4664   DAD

parallel lines   220   Lines in the same plane that never cross.

parallelogram   226   A quadrilateral whose opposite sides are parallel.

parentheses   28   Used to show which operation to do first in an equation.

$$6 + (4 - 2) = 6 + 2 = 8$$

pentagon   224   A polygon with five sides.

pentominoes   236   Shapes created by arranging five squares.

perimeter   238   The distance around a figure. To find the perimeter, add the lengths of the sides.

period   2   Each group of three digits in a number. Periods are separated by commas.

430,562,784

perpendicular lines   220
Two intersecting lines that form square corners are perpendicular.

pictograph   120   A graph that uses pictures to show numbers of objects.

pint (pt)   192   A unit of capacity in the customary system of measurement.

1 pint = 2 cups

place value   2   A system for writing numbers. In this system, the position of a digit determines its value.

plane   218   A never-ending, flat surface.

P.M.   250   The hours from noon to midnight.

point   218   An exact location in space.

polygon   224   A closed plane figure with straight sides that do not cross.

pound (lb)   194   A unit of weight in the customary system of measurement.

1 pound = 16 ounces

prime numbers   81   A number with only two factors, one and itself.

probability   270   The chance that something will happen. Probability is given as a fraction. The numerator names the number of ways the event can occur. The denominator names the total number of possible outcomes.

The probability of spinning red is $\frac{2}{6}$.

pyramid   242   A solid figure with triangular faces and a base that is a polygon.

Q

quadrilateral   224
A polygon with four sides.

quart (qt)   192   A unit of capacity in the customary system of measurement.

1 quart = 2 pints

quotient   150   The answer to a division problem.

R

random   270   Each choice is equally likely.

range   108   The difference between the greatest and the least number in a set of numbers.

ray   222   A part of a line with only one endpoint.

rectangle   226
A parallelogram with four right angles.

rectangular prism   242
A solid figure that has six rectangular faces like a box.

reflection   234   Also called a flip; a motion that makes a figure face in the opposite direction.

regroup   36   To exchange equal amounts when computing.

12 = 1 ten and 2 ones

remainder   140   The amount left after dividing one whole number by another. The remainder is always less than the divisor.

$$\begin{array}{r} 2\ R1 \\ 4\overline{)9} \\ 8 \\ \hline 1 \end{array}$$ remainder

rename   36   See *regroup.*

rhombus 226 A quadrilateral with all four sides the same length.

right angle 222 An angle that exactly fits the corner of a square.

rotation 234 Also called a turn; the motion of turning a figure around a point.

rounded number 16 A number that tells about how many. 28 rounded to the nearest ten is 30.

rounding numbers 20 Approximating the value of a number relative to a stated place value.

> **28 rounded to the nearest 10 is 30.**
> **22 rounded to the nearest 10 is 20.**

S

sieve 81 A tool used to sift or funnel items.

similar figures 228 Two or more figures that have the same shape but might differ in size.

simplest form of a fraction 260 A fraction is in simplest form when the only common factor of the numerator and the denominator is one.

$$\frac{4}{12} = \frac{1}{3}$$

slide 234 See *translation*

sphere 242 A solid figure that is shaped like a round ball. A sphere has no faces, edges, or vertices.

square 226 A rectangle with four congruent sides.

square pyramid 242 A solid figure with a square base and 4 triangular sides.

standard form 2 The way a number is usually written. The standard form of thirty-six is 36.

statistics 105 Statistics involves collecting, organizing, interpreting, and displaying data.

stem-and-leaf plot 118 A display which lists data by place value. The ones digits are the leaves, and the tens digits are the stems.

subtraction 42 The operation used to take away from a group or to compare groups.

> **9 − 4 = 5**

sum 28 The answer to an addition problem. In the addition sentence 6 + 7 = 13, 13 is the sum.

T

tally chart 106 A display in which a mark is made in a table to show each time a piece of data occurs.

ton 194 A unit of weight in the customary system.

> **1 ton = 2,000 pounds**

transformation 234 Moving a figure in geometry; a reflection, rotation or a translation.

translation 234 Also called a slide; the motion of sliding a figure in a straight line.

trapezoid 226 A quadrilateral with exactly one pair of parallel sides.

tree diagram 274 A diagram that uses branches to show all possible outcomes.

triangle 224 A polygon with three sides.

triangular prism 242 A solid figure that has two congruent triangles for bases. The other three faces of a triangular prism are rectangles.

turn   **224**   See *rotation*.

V

variables   **124**   Letters or symbols that stand for actual numbers.

Venn diagram   **318**
Overlapping circles enclosed in a rectangle that show how objects are classified.

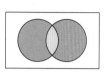

vertex of a polygon   **224, 242**
A point where two sides of the polygon meet.

vertex

volume   **244**   The total amount of space a figure takes up.

W

weight   **202**   The measure of the heaviness of an object.

X

x-axis   **126**   The horizontal number line on the grid.

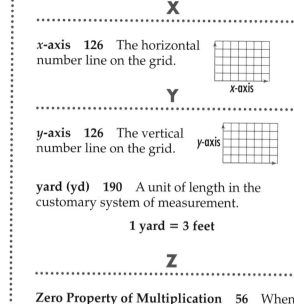

x-axis

Y

y-axis   **126**   The vertical number line on the grid.

y-axis

yard (yd)   **190**   A unit of length in the customary system of measurement.

1 yard = 3 feet

Z

Zero Property of Multiplication   **56**   When 0 is a factor, the product is 0.

INDEX